Standards-Based Math

3-4

Written by
Kim Cernek and Denise Skomer

Editor: Carla Hamaguchi
Illustrator: Jenny Campbell
Designer/Production: Moonhee Pak/Shirley A. Cabrera
Cover Designer: Barbara Peterson
Art Director: Tom Cochrane
Project Director: Carolea Williams

Table of Contents

Introduction

Each book in the *Power Practice*™ series contains dozens of ready-to-use activity pages to provide students with skill practice. The fun activities can be used to supplement and enhance what you are already teaching in your classroom. Give an activity page to students as independent class work, or send the pages home as homework to reinforce skills taught in class. An answer key is provided at the end of each book for quick reference.

Standards-Based Math 3–4 provides activities that will directly assist students in practicing and reinforcing math skills such as place value, addition, subtraction, multiplication, division, word problems, fractions, measurement, probability, and more!

The book is based on the National Council of Teachers of Mathematics (NCTM) standards and is divided into five main sections: Number and Operations, Algebra, Geometry, Measurement, and Data Analysis and Probability. You'll find several activity pages in each section that will motivate students and reinforce the math skills.

Use these ready-to-go activities to "recharge" skill review and give students the power to succeed!

Name _____ Date _____

Place Value

NUMBER AND OPERATIONS

millions	hundred thousands	ten thousands	thousands	hundreds	tens	ones
6,	7	8	9,	5	3	2

What place is the 6 in?

1 456

_____

2 6,545,000

_____

3 5,634,000

_____

4 546,345

_____

5 634,998

_____

6 4,536

_____

7 3,654

_____

8 4,365,777

_____

9 4,361

_____

10 6,423

_____

11 60

_____

12 464,999

_____

Addition—No Regrouping

NUMBER AND OPERATIONS

	Add ones.	Add tens.
	25	25
	+ 13	+ 13
	8	38

①
15	24	38	52	28
+ 4	+ 15	+ 21	+ 14	+ 11

②
72	53	36	44	20
+ 17	+ 15	+ 22	+ 35	+ 34

③
31	64	26	33	54
+ 22	+ 22	+ 53	+ 33	+ 12

④
56	25	73	41	18
+ 11	+ 52	+ 13	+ 26	+ 31

⑤
43	27	23	26	45
+ 34	+ 52	+ 45	+ 23	+ 14

Standards-Based Math • 3–4 © 2004 Creative Teaching Press

Name _____   Date _____

Score!

25
+ 73

92
+ 57

110
+ 40

63
+ 35

35
+ 64

43
+ 54

31
+ 65

85
+ 12

32
+ 42

42
+ 33

Even | Odd

Is the sum of each problem even or odd? Place a tally mark on the scoreboard for each sum.

Standards-Based Math • 3-4 © 2004 Creative Teaching Press

Addition—Regrouping

NUMBER AND OPERATIONS

Add ones.	Regroup.		Add tens.	Regroup.
	$\begin{array}{r} 1 \\ 25 \\ +\ 87 \\ \hline 2 \end{array}$			$\begin{array}{r} 1 \\ 25 \\ +\ 87 \\ \hline 112 \end{array}$

1
$\begin{array}{r} 75 \\ +\ 95 \\ \hline \end{array}$ $\begin{array}{r} 84 \\ +\ 69 \\ \hline \end{array}$ $\begin{array}{r} 78 \\ +\ 46 \\ \hline \end{array}$ $\begin{array}{r} 65 \\ +\ 86 \\ \hline \end{array}$ $\begin{array}{r} 88 \\ +\ 55 \\ \hline \end{array}$

2
$\begin{array}{r} 94 \\ +\ 27 \\ \hline \end{array}$ $\begin{array}{r} 88 \\ +\ 34 \\ \hline \end{array}$ $\begin{array}{r} 68 \\ +\ 99 \\ \hline \end{array}$ $\begin{array}{r} 77 \\ +\ 36 \\ \hline \end{array}$ $\begin{array}{r} 67 \\ +\ 94 \\ \hline \end{array}$

3
$\begin{array}{r} 168 \\ +\ 58 \\ \hline \end{array}$ $\begin{array}{r} 168 \\ +\ 24 \\ \hline \end{array}$ $\begin{array}{r} 146 \\ +\ 27 \\ \hline \end{array}$ $\begin{array}{r} 177 \\ +\ 39 \\ \hline \end{array}$ $\begin{array}{r} 135 \\ +\ 75 \\ \hline \end{array}$

4
$\begin{array}{r} 148 \\ +\ 99 \\ \hline \end{array}$ $\begin{array}{r} 158 \\ +\ 28 \\ \hline \end{array}$ $\begin{array}{r} 166 \\ +\ 46 \\ \hline \end{array}$ $\begin{array}{r} 155 \\ +\ 75 \\ \hline \end{array}$ $\begin{array}{r} 208 \\ +\ 93 \\ \hline \end{array}$

5
$\begin{array}{r} 164 \\ +\ 48 \\ \hline \end{array}$ $\begin{array}{r} 207 \\ +\ 89 \\ \hline \end{array}$ $\begin{array}{r} 148 \\ +\ 65 \\ \hline \end{array}$ $\begin{array}{r} 136 \\ +\ 46 \\ \hline \end{array}$ $\begin{array}{r} 226 \\ +\ 90 \\ \hline \end{array}$

Standards-Based Math • 3–4 © 2004 Creative Teaching Press

Name _____ Date _____

Three Addends

NUMBER AND OPERATIONS

1
```
  17        24        18        15        18
  14        12        13        12        18
+  5      +  9      + 16      + 15      + 16
```

2
```
  20        32        29        33        44
  17        26        39        16        23
+ 15      + 14      + 19      + 22      + 34
```

3
```
  22        34        32        35        39
  15        33        43        24        24
+ 26      + 19      + 15      + 35      + 46
```

4
```
  19        30        42        12        48
  23        26        16        27        34
+ 15      + 14      + 19      + 41      + 23
```

Name _____ Date _____

Bowling for Numbers

NUMBER AND OPERATIONS

Color a combination of pins that totals 100 points. Write an equation for each picture.

1

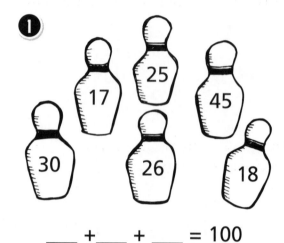

___ +___ + ___ = 100

4

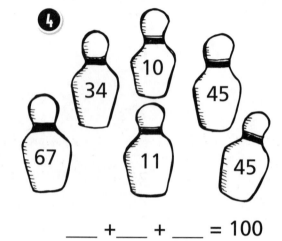

___ +___ + ___ = 100

2

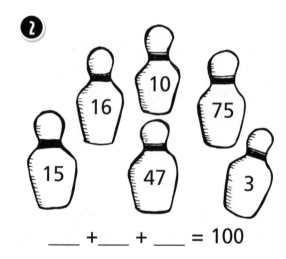

___ +___ + ___ = 100

5

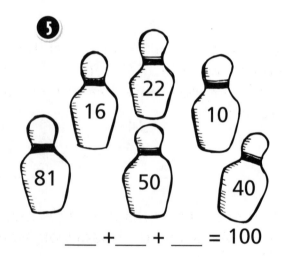

___ +___ + ___ = 100

3

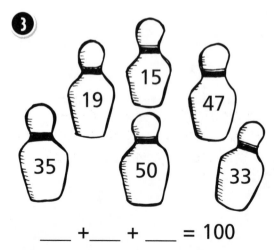

___ +___ + ___ = 100

6

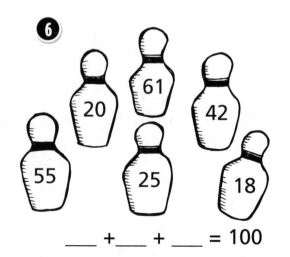

___ +___ + ___ = 100

Word Problems 1

NUMBER AND OPERATIONS

1 Lindsay scored 32 points today, 24 points yesterday, and 26 points the day before that. How many points has Lindsay scored in her last three basketball games?

2 There are 27 fourth-grade students in Mrs. Mayefsky's class, 32 in Mrs. Albright's class, and 28 in Mr. Pischke's class. How many fourth-grade students are there in all?

3 The class that sells 100 tickets to the dance earns a free pizza party. Miss Krema's class sold 16 tickets on Monday, 29 tickets on Wednesday, and 50 tickets on Friday. How many more tickets do they need to sell to earn the pizza party?

4 Each night Marcus reads 5 pages more of his novel than he did the night before. Marcus read 12 pages on Monday. How many pages will he have read by Thursday?

5 Emily has twin sisters. Emily buys double of everything for them for their birthday. She bought each sister a $6.00 pair of sunglasses, a $1.00 ball, and a $2.00 book. How much money did Emily spend on birthday presents in all?

Name _____ Date _____

Subtraction–No Regrouping

NUMBER AND OPERATIONS

	Subtract ones.	Subtract tens.
	$\begin{array}{r} 25 \\ -\ 13 \\ \hline 2 \end{array}$	$\begin{array}{r} 25 \\ -\ 13 \\ \hline 12 \end{array}$

1
$\begin{array}{r} 46 \\ -\ 16 \\ \hline \end{array}$
$\begin{array}{r} 49 \\ -\ 11 \\ \hline \end{array}$
$\begin{array}{r} 35 \\ -\ 12 \\ \hline \end{array}$
$\begin{array}{r} 26 \\ -\ 23 \\ \hline \end{array}$
$\begin{array}{r} 55 \\ -\ 22 \\ \hline \end{array}$

2
$\begin{array}{r} 27 \\ -\ 7 \\ \hline \end{array}$
$\begin{array}{r} 52 \\ -\ 20 \\ \hline \end{array}$
$\begin{array}{r} 29 \\ -\ 25 \\ \hline \end{array}$
$\begin{array}{r} 75 \\ -\ 23 \\ \hline \end{array}$
$\begin{array}{r} 126 \\ -\ 24 \\ \hline \end{array}$

3
$\begin{array}{r} 97 \\ -\ 55 \\ \hline \end{array}$
$\begin{array}{r} 82 \\ -\ 40 \\ \hline \end{array}$
$\begin{array}{r} 66 \\ -\ 21 \\ \hline \end{array}$
$\begin{array}{r} 28 \\ -\ 14 \\ \hline \end{array}$
$\begin{array}{r} 157 \\ -\ 52 \\ \hline \end{array}$

4
$\begin{array}{r} 256 \\ -\ 33 \\ \hline \end{array}$
$\begin{array}{r} 476 \\ -\ 52 \\ \hline \end{array}$
$\begin{array}{r} 555 \\ -\ 43 \\ \hline \end{array}$
$\begin{array}{r} 676 \\ -\ 54 \\ \hline \end{array}$
$\begin{array}{r} 299 \\ -\ 67 \\ \hline \end{array}$

5
$\begin{array}{r} 776 \\ -\ 34 \\ \hline \end{array}$
$\begin{array}{r} 587 \\ -\ 63 \\ \hline \end{array}$
$\begin{array}{r} 877 \\ -\ 61 \\ \hline \end{array}$
$\begin{array}{r} 584 \\ -\ 42 \\ \hline \end{array}$
$\begin{array}{r} 967 \\ -\ 57 \\ \hline \end{array}$

Standards-Based Math • 3–4 © 2004 Creative Teaching Press

Take a Moonwalk

Number and Operations

Help the astronaut get back to the shuttle. He can only step on moon rocks that are marked with problems that have correct answers. Cross out the problems with incorrect answers to mark the path.

Subtraction–Regrouping

Number and Operations

Regroup. ten = 10 ones	Subtract ones.	Regroup.	Subtract tens.
$2\overset{2}{3}\overset{1}{5}$ $-\ 47$	$2\overset{2}{3}\overset{1}{5}$ $-\ 47$ 8	$\overset{1}{2}\overset{1}{3}\overset{2}{\underset{1}{5}}$ $-\ 47$ 8	$\overset{1}{2}\overset{1}{3}\overset{2}{\underset{1}{5}}$ $-\ 47$ 188

1.

$$54 - 35 \qquad 73 - 28 \qquad 84 - 37 \qquad 26 - 18 \qquad 62 - 39$$

2.

$$136 - 47 \qquad 176 - 87 \qquad 88 - 39 \qquad 74 - 46 \qquad 95 - 36$$

3.

$$223 - 45 \qquad 456 - 87 \qquad 185 - 99 \qquad 85 - 38 \qquad 44 - 15$$

4.

$$161 - 48 \qquad 125 - 67 \qquad 233 - 77 \qquad 615 - 86 \qquad 30 - 14$$

5.

$$236 - 59 \qquad 94 - 68 \qquad 62 - 19 \qquad 333 - 88 \qquad 545 - 97$$

Standards-Based Math • 3–4 © 2004 Creative Teaching Press

Number, Please

Number and Operations

Solve each problem. Use the chart to see which town you are calling.

413
− 85

This is the
area code

for _____.

614
−147

This is the
area code

for _____.

511
−274

This is the
area code

for _____.

442
−114

This is the
area code

for _____.

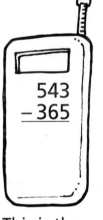

543
−365

This is the
area code

for _____.

347
−169

This is the
area code

for _____.

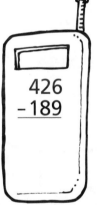

426
−189

This is the
area code

for _____.

522
− 55

This is the
area code

for _____.

Area Code Directory

178 Jackson	237 Linden
328 Zanville	467 Carlstown

Standards-Based Math • 3–4 © 2004 Creative Teaching Press

Name _____   Date _____

Subtraction Soup

Number and Operations

Help the chef figure out which ingredient is missing from each pot of subtraction soup.

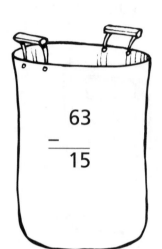
$$\begin{array}{r} 63 \\ -\ 15 \\ \hline \end{array}$$

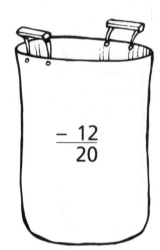
$$\begin{array}{r} -\ 12 \\ 20 \\ \hline \end{array}$$

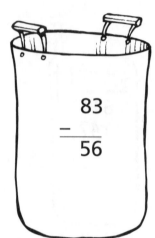
$$\begin{array}{r} 83 \\ -\ \ \ \\ \hline 56 \end{array}$$

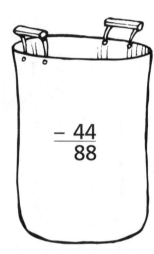
$$\begin{array}{r} -\ 44 \\ 88 \\ \hline \end{array}$$

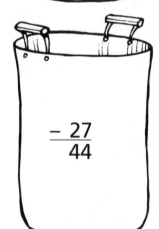
$$\begin{array}{r} -\ 27 \\ 44 \\ \hline \end{array}$$

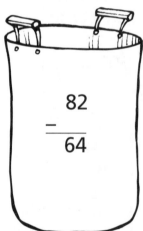
$$\begin{array}{r} 82 \\ -\ \ \ \\ \hline 64 \end{array}$$

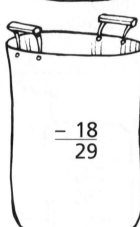
$$\begin{array}{r} -\ 18 \\ 29 \\ \hline \end{array}$$

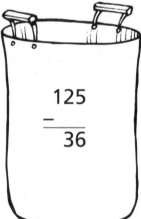
$$\begin{array}{r} 125 \\ -\ \ \ \\ \hline 36 \end{array}$$

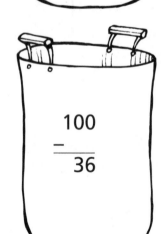
$$\begin{array}{r} 100 \\ -\ \ \ \\ \hline 36 \end{array}$$

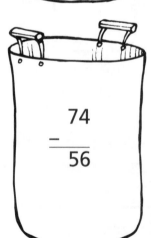
$$\begin{array}{r} 74 \\ -\ \ \ \\ \hline 56 \end{array}$$

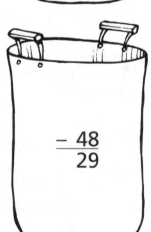
$$\begin{array}{r} -\ 48 \\ 29 \\ \hline \end{array}$$

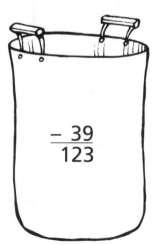
$$\begin{array}{r} -\ 39 \\ 123 \\ \hline \end{array}$$

Standards-Based Math • 3–4 © 2004 Creative Teaching Press

Name _____ Date _____

Word Problems 2

NUMBER AND OPERATIONS

1 There are enough seats for 120 people at the high school basketball game. 178 people arrive at the ticket booth. How many of them will have to stand to watch the game?

2 William needs 13 more quarters to complete his collection of coins from all 50 states. How many does he have already?

3 There is space for 500 photos in Sarah's album. She has 180 pictures in it so far. How many more photos can she add?

4 The first person to sell 80 magazine subscriptions wins a free backpack. Sal needs to sell 17 more. Sal has sold 3 more than Kelly. How many magazine subscriptions has Kelly sold?

5 José bought a pack of 100 sheets of paper. He used 16 to write his report and gave some to his sister. José has 32 pieces of paper left. How many did he give to his sister?

Multiplication 0–5

NUMBER AND OPERATIONS

1 $0 \times 2 =$ $3 \times 2 =$ $4 \times 6 =$ $5 \times 8 =$

2 $3 \times 1 =$ $3 \times 3 =$ $1 \times 9 =$ $2 \times 7 =$

3 $4 \times 4 =$ $3 \times 7 =$ $5 \times 3 =$ $4 \times 2 =$

4 $2 \times 6 =$ $3 \times 5 =$ $4 \times 11 =$ $5 \times 9 =$

5 $2 \times 4 =$ $3 \times 9 =$ $4 \times 7 =$ $5 \times 4 =$

6 $8 \times 2 =$ $8 \times 3 =$ $10 \times 4 =$ $11 \times 5 =$

7 $6 \times 3 =$ $0 \times 6 =$ $2 \times 8 =$ $3 \times 10 =$

8 $4 \times 7 =$ $2 \times 12 =$ $5 \times 5 =$ $3 \times 7 =$

Standards-Based Math • 3–4 © 2004 Creative Teaching Press

Multiplication 6–12

NUMBER AND OPERATIONS

1 $6 \times 4 =$ $7 \times 2 =$ $8 \times 3 =$ $9 \times 5 =$

2 $6 \times 3 =$ $6 \times 9 =$ $7 \times 4 =$ $8 \times 4 =$

3 $9 \times 7 =$ $7 \times 8 =$ $6 \times 5 =$ $7 \times 6 =$

4 $8 \times 2 =$ $9 \times 4 =$ $8 \times 6 =$ $6 \times 6 =$

5 $8 \times 7 =$ $10 \times 8 =$ $9 \times 9 =$ $9 \times 11 =$

6 $10 \times 6 =$ $9 \times 7 =$ $12 \times 8 =$ $10 \times 9 =$

7 $12 \times 7 =$ $8 \times 8 =$ $7 \times 7 =$ $9 \times 12 =$

8 $6 \times 11 =$ $9 \times 3 =$ $11 \times 11 =$ $12 \times 12 =$

Name _____   Date _____

 # Multiplication–No Regrouping

Number and Operations

Multiply each digit of the top number by the bottom number.

Start with the ones place.	Then, multiply the tens place.
13 × 2 —— 6	13 × 2 —— 26

1
| 12
 × 4 | 11
 × 2 | 13
 × 3 | 21
 × 3 | 41
 × 2 |

2
| 11
 × 4 | 22
 × 3 | 10
 × 7 | 11
 × 6 | 34
 × 2 |

3
| 44
 × 2 | 11
 × 8 | 10
 × 2 | 13
 × 2 | 43
 × 2 |

4
| 10
 × 9 | 21
 × 4 | 32
 × 2 | 24
 × 2 | 44
 × 1 |

5
| 12
 × 3 | 14
 × 2 | 31
 × 3 | 11
 × 9 | 32
 × 3 |

Standards-Based Math • 3–4 © 2004 Creative Teaching Press

Multiplication—Regrouping

NUMBER AND OPERATIONS

Multiply each digit of the top number by the bottom number.
Regroup if the product is 10 or above.
Start with the ones place.   Then, multiply the tens place, and
add the regrouped portion.

$$\begin{array}{r} ^1 15 \\ \times \quad 2 \\ \hline 0 \end{array} \qquad \begin{array}{r} ^1 15 \\ \times \quad 2 \\ \hline 30 \end{array}$$

①
$$\begin{array}{r} 13 \\ \times \ 4 \\ \hline \end{array} \qquad \begin{array}{r} 17 \\ \times \ 2 \\ \hline \end{array} \qquad \begin{array}{r} 13 \\ \times \ 5 \\ \hline \end{array} \qquad \begin{array}{r} 24 \\ \times \ 3 \\ \hline \end{array} \qquad \begin{array}{r} 25 \\ \times \ 2 \\ \hline \end{array}$$

②
$$\begin{array}{r} 18 \\ \times \ 4 \\ \hline \end{array} \qquad \begin{array}{r} 25 \\ \times \ 3 \\ \hline \end{array} \qquad \begin{array}{r} 22 \\ \times \ 7 \\ \hline \end{array} \qquad \begin{array}{r} 18 \\ \times \ 6 \\ \hline \end{array} \qquad \begin{array}{r} 35 \\ \times \ 2 \\ \hline \end{array}$$

③
$$\begin{array}{r} 44 \\ \times \ 5 \\ \hline \end{array} \qquad \begin{array}{r} 16 \\ \times \ 8 \\ \hline \end{array} \qquad \begin{array}{r} 29 \\ \times \ 3 \\ \hline \end{array} \qquad \begin{array}{r} 26 \\ \times \ 4 \\ \hline \end{array} \qquad \begin{array}{r} 45 \\ \times \ 2 \\ \hline \end{array}$$

④
$$\begin{array}{r} 15 \\ \times \ 9 \\ \hline \end{array} \qquad \begin{array}{r} 28 \\ \times \ 4 \\ \hline \end{array} \qquad \begin{array}{r} 48 \\ \times \ 4 \\ \hline \end{array} \qquad \begin{array}{r} 63 \\ \times \ 6 \\ \hline \end{array} \qquad \begin{array}{r} 55 \\ \times \ 2 \\ \hline \end{array}$$

⑤
$$\begin{array}{r} 23 \\ \times \ 8 \\ \hline \end{array} \qquad \begin{array}{r} 34 \\ \times \ 7 \\ \hline \end{array} \qquad \begin{array}{r} 47 \\ \times \ 8 \\ \hline \end{array} \qquad \begin{array}{r} 54 \\ \times \ 9 \\ \hline \end{array} \qquad \begin{array}{r} 77 \\ \times \ 5 \\ \hline \end{array}$$

Standards-Based Math • 3–4 © 2004 Creative Teaching Press

Two-Digit Multiplication—No Regrouping

NUMBER AND OPERATIONS

Multiply each digit of the top number by the ones digit in the bottom number.

$$\begin{array}{r} 22 \\ \times\ 12 \\ \hline 44 \end{array}$$

Then place a 0 in the ones place. Multiply each digit of the top number by the tens digit in the bottom number.

$$\begin{array}{r} 22 \\ \times\ 12 \\ \hline 44 \\ 220 \end{array}$$

Then add the products.

$$\begin{array}{r} 22 \\ \times\ 12 \\ \hline 44 \\ +\ 220 \\ \hline 264 \end{array}$$

1

$$\begin{array}{r} 12 \\ \times\ 11 \end{array} \qquad \begin{array}{r} 11 \\ \times\ 13 \end{array} \qquad \begin{array}{r} 13 \\ \times\ 13 \end{array} \qquad \begin{array}{r} 21 \\ \times\ 13 \end{array} \qquad \begin{array}{r} 42 \\ \times\ 12 \end{array}$$

2

$$\begin{array}{r} 11 \\ \times\ 24 \end{array} \qquad \begin{array}{r} 22 \\ \times\ 23 \end{array} \qquad \begin{array}{r} 10 \\ \times\ 37 \end{array} \qquad \begin{array}{r} 31 \\ \times\ 21 \end{array} \qquad \begin{array}{r} 34 \\ \times\ 22 \end{array}$$

3

$$\begin{array}{r} 43 \\ \times\ 12 \end{array} \qquad \begin{array}{r} 11 \\ \times\ 28 \end{array} \qquad \begin{array}{r} 20 \\ \times\ 42 \end{array} \qquad \begin{array}{r} 13 \\ \times\ 31 \end{array} \qquad \begin{array}{r} 43 \\ \times\ 22 \end{array}$$

4

$$\begin{array}{r} 55 \\ \times\ 11 \end{array} \qquad \begin{array}{r} 20 \\ \times\ 10 \end{array} \qquad \begin{array}{r} 32 \\ \times\ 32 \end{array} \qquad \begin{array}{r} 25 \\ \times\ 21 \end{array} \qquad \begin{array}{r} 44 \\ \times\ 21 \end{array}$$

Standards-Based Math • 3-4 © 2004 Creative Teaching Press

Name _____ Date _____

Dragon Digits

NUMBER AND OPERATIONS

Color the Chinese Dragon that has the products in order from least to greatest.

1

$$44 \times 22$$ $$23 \times 12$$ $$16 \times 11$$

2

$$14 \times 10$$ $$33 \times 32$$ $$41 \times 13$$

3

$$32 \times 23$$ $$15 \times 11$$ $$41 \times 11$$

4

$$14 \times 12$$ $$23 \times 13$$ $$43 \times 21$$

Standards-Based Math • 3–4 © 2004 Creative Teaching Press

23

Two-Digit Multiplication–Regrouping

NUMBER AND OPERATIONS

1
$$\begin{array}{r} 13 \\ \times\ 14 \\ \hline \end{array}$$
$$\begin{array}{r} 16 \\ \times\ 12 \\ \hline \end{array}$$
$$\begin{array}{r} 15 \\ \times\ 13 \\ \hline \end{array}$$
$$\begin{array}{r} 21 \\ \times\ 13 \\ \hline \end{array}$$
$$\begin{array}{r} 45 \\ \times\ 12 \\ \hline \end{array}$$

2
$$\begin{array}{r} 18 \\ \times\ 24 \\ \hline \end{array}$$
$$\begin{array}{r} 28 \\ \times\ 23 \\ \hline \end{array}$$
$$\begin{array}{r} 16 \\ \times\ 37 \\ \hline \end{array}$$
$$\begin{array}{r} 37 \\ \times\ 46 \\ \hline \end{array}$$
$$\begin{array}{r} 37 \\ \times\ 22 \\ \hline \end{array}$$

3
$$\begin{array}{r} 43 \\ \times\ 16 \\ \hline \end{array}$$
$$\begin{array}{r} 29 \\ \times\ 28 \\ \hline \end{array}$$
$$\begin{array}{r} 35 \\ \times\ 42 \\ \hline \end{array}$$
$$\begin{array}{r} 13 \\ \times\ 35 \\ \hline \end{array}$$
$$\begin{array}{r} 46 \\ \times\ 42 \\ \hline \end{array}$$

4
$$\begin{array}{r} 55 \\ \times\ 39 \\ \hline \end{array}$$
$$\begin{array}{r} 26 \\ \times\ 46 \\ \hline \end{array}$$
$$\begin{array}{r} 38 \\ \times\ 32 \\ \hline \end{array}$$
$$\begin{array}{r} 25 \\ \times\ 28 \\ \hline \end{array}$$
$$\begin{array}{r} 48 \\ \times\ 42 \\ \hline \end{array}$$

5
$$\begin{array}{r} 66 \\ \times\ 12 \\ \hline \end{array}$$
$$\begin{array}{r} 73 \\ \times\ 24 \\ \hline \end{array}$$
$$\begin{array}{r} 36 \\ \times\ 42 \\ \hline \end{array}$$
$$\begin{array}{r} 74 \\ \times\ 28 \\ \hline \end{array}$$
$$\begin{array}{r} 16 \\ \times\ 88 \\ \hline \end{array}$$

Standards-Based Math • 3-4 © 2004 Creative Teaching Press

Multiplying for Medals

Number and Operations

Solve the problems to learn whether each medal for the Greek Olympics is bronze, silver. or gold.

| Bronze = 25–49 | Silver = 50–74 | Gold = 75–100 |

1
$$\begin{array}{r} 23 \\ \times\ \ 4 \\ \hline \end{array}$$

$$\begin{array}{r} 46 \\ \times\ \ 2 \\ \hline \end{array}$$

$$\begin{array}{r} 36 \\ \times\ \ 2 \\ \hline \end{array}$$

_____   _____   _____

2
$$\begin{array}{r} 35 \\ \times\ \ 2 \\ \hline \end{array}$$
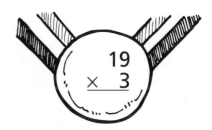
$$\begin{array}{r} 19 \\ \times\ \ 3 \\ \hline \end{array}$$

_____   _____

3
$$\begin{array}{r} 12 \\ \times\ \ 4 \\ \hline \end{array}$$
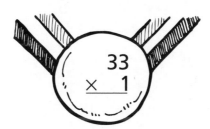
$$\begin{array}{r} 33 \\ \times\ \ 1 \\ \hline \end{array}$$

$$\begin{array}{r} 13 \\ \times\ \ 5 \\ \hline \end{array}$$

_____   _____

4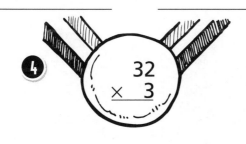
$$\begin{array}{r} 32 \\ \times\ \ 3 \\ \hline \end{array}$$

$$\begin{array}{r} 20 \\ \times\ \ 5 \\ \hline \end{array}$$

_____   _____

Three-Digit Multiplication—Regrouping

Number and Operations

1

$$\begin{array}{r} 215 \\ \times\ 4 \\ \hline \end{array}$$
$$\begin{array}{r} 107 \\ \times\ 4 \\ \hline \end{array}$$
$$\begin{array}{r} 313 \\ \times\ 5 \\ \hline \end{array}$$
$$\begin{array}{r} 246 \\ \times\ 3 \\ \hline \end{array}$$
$$\begin{array}{r} 415 \\ \times\ 7 \\ \hline \end{array}$$

2

$$\begin{array}{r} 518 \\ \times\ 3 \\ \hline \end{array}$$
$$\begin{array}{r} 248 \\ \times\ 6 \\ \hline \end{array}$$
$$\begin{array}{r} 193 \\ \times\ 5 \\ \hline \end{array}$$
$$\begin{array}{r} 361 \\ \times\ 4 \\ \hline \end{array}$$
$$\begin{array}{r} 234 \\ \times\ 7 \\ \hline \end{array}$$

3

$$\begin{array}{r} 423 \\ \times\ 6 \\ \hline \end{array}$$
$$\begin{array}{r} 199 \\ \times\ 2 \\ \hline \end{array}$$
$$\begin{array}{r} 208 \\ \times\ 6 \\ \hline \end{array}$$
$$\begin{array}{r} 413 \\ \times\ 3 \\ \hline \end{array}$$
$$\begin{array}{r} 243 \\ \times\ 8 \\ \hline \end{array}$$

4

$$\begin{array}{r} 525 \\ \times\ 3 \\ \hline \end{array}$$
$$\begin{array}{r} 236 \\ \times\ 6 \\ \hline \end{array}$$
$$\begin{array}{r} 132 \\ \times\ 8 \\ \hline \end{array}$$
$$\begin{array}{r} 225 \\ \times\ 4 \\ \hline \end{array}$$
$$\begin{array}{r} 428 \\ \times\ 5 \\ \hline \end{array}$$

Standards-Based Math • 3–4 © 2004 Creative Teaching Press

Name _____ Date _____

Four-Digit Multiplication–Regrouping

NUMBER AND OPERATIONS

1
$$2415 \times 3$$
$$1907 \times 2$$
$$3213 \times 4$$
$$2646 \times 5$$
$$4815 \times 3$$

2
$$5638 \times 2$$
$$6248 \times 4$$
$$7811 \times 3$$
$$4361 \times 4$$
$$5534 \times 2$$

3
$$4237 \times 3$$
$$1959 \times 8$$
$$6058 \times 6$$
$$4583 \times 2$$
$$1243 \times 8$$

4
$$5125 \times 3$$
$$7236 \times 3$$
$$1324 \times 9$$
$$3295 \times 4$$
$$2424 \times 5$$

Word Problems 3

NUMBER AND OPERATIONS

1 Orlando will give each child at his party 4 balloons. There are 16 children coming to his party. How many balloons will Orlando give away?

2 Karen gets 5¢ for each can she collects. How much will she receive for 35 cans?

3 Suki's mom has two brothers and one sister. Her uncles both have 4 kids each. Her aunt has 2 kids. How many cousins does Suki have?

4 Jacob can plant 5 daffodil bulbs in each square foot of his garden. Jacob's garden is 16 square feet. How many bulbs should Jacob buy?

5 Beta's scout troop is making kites. Each kite needs 25 yards of string. There are 12 people in Beta's troop. String comes in rolls of 100 yards. How many rolls of string will Beta's troop need to buy?

Standards-Based Math • 3–4 © 2004 Creative Teaching Press

Division 0-5

Number and Operations

1 $1\overline{)7}$ $2\overline{)8}$ $2\overline{)4}$ $1\overline{)6}$ $1\overline{)3}$

2 $2\overline{)10}$ $1\overline{)1}$ $0\overline{)6}$ $1\overline{)4}$ $0\overline{)10}$

3 $1\overline{)3}$ $2\overline{)12}$ $2\overline{)6}$ $1\overline{)2}$ $1\overline{)5}$

4 $4\overline{)24}$ $5\overline{)5}$ $1\overline{)7}$ $3\overline{)12}$ $5\overline{)10}$

5 $5\overline{)15}$ $3\overline{)27}$ $4\overline{)32}$ $3\overline{)9}$ $5\overline{)35}$

6 $4\overline{)8}$ $5\overline{)20}$ $3\overline{)15}$ $4\overline{)16}$ $3\overline{)9}$

Skyscraper Stories

NUMBER AND OPERATIONS

The Sears Tower in Chicago, Illinois, is the tallest building in North America. Solve the problems. Then add the quotients to figure out how many stories tall it is.

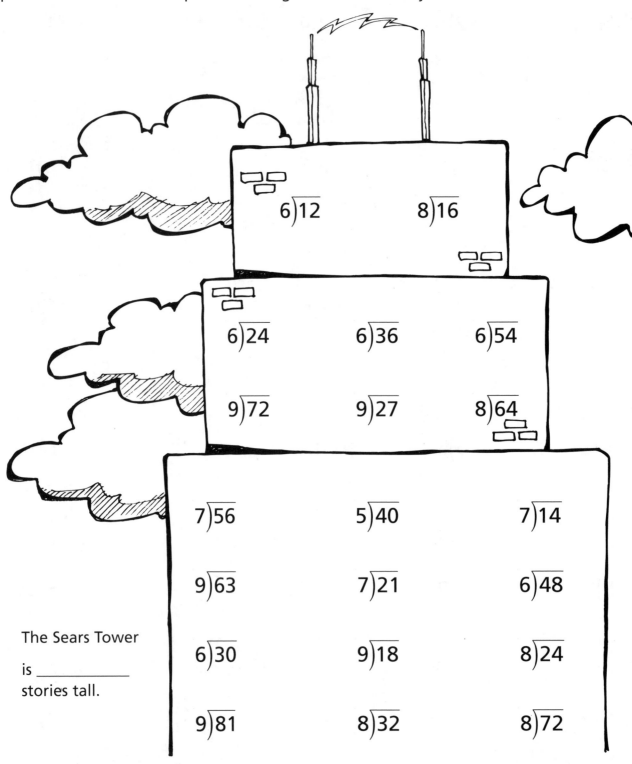

$6\overline{)12}$ $8\overline{)16}$

$6\overline{)24}$ $6\overline{)36}$ $6\overline{)54}$

$9\overline{)72}$ $9\overline{)27}$ $8\overline{)64}$

$7\overline{)56}$ $5\overline{)40}$ $7\overline{)14}$

$9\overline{)63}$ $7\overline{)21}$ $6\overline{)48}$

The Sears Tower

is _____

stories tall.

$6\overline{)30}$ $9\overline{)18}$ $8\overline{)24}$

$9\overline{)81}$ $8\overline{)32}$ $8\overline{)72}$

Two-Digit Division—No Remainders

Number and Operations

1 $2\overline{)46}$ $3\overline{)39}$ $6\overline{)66}$ $4\overline{)48}$

2 $2\overline{)28}$ $4\overline{)84}$ $5\overline{)55}$ $3\overline{)69}$

3 $4\overline{)64}$ $5\overline{)65}$ $2\overline{)38}$ $6\overline{)78}$

4 $3\overline{)54}$ $7\overline{)84}$ $4\overline{)68}$ $2\overline{)78}$

5 $4\overline{)96}$ $3\overline{)87}$ $5\overline{)75}$ $3\overline{)78}$

6 $6\overline{)84}$ $7\overline{)98}$ $8\overline{)88}$ $5\overline{)85}$

Standards-Based Math • 3–4 © 2004 Creative Teaching Press

Three-Digit Division—No Remainders

NUMBER AND OPERATIONS

❶ $2\overline{)426}$ $3\overline{)639}$ $4\overline{)488}$ $2\overline{)484}$

❷ $2\overline{)346}$ $3\overline{)396}$ $4\overline{)104}$ $5\overline{)110}$

❸ $4\overline{)112}$ $3\overline{)108}$ $6\overline{)108}$ $7\overline{)147}$

❹ $4\overline{)232}$ $5\overline{)305}$ $8\overline{)336}$ $7\overline{)448}$

❺ $9\overline{)648}$ $4\overline{)384}$ $9\overline{)756}$ $8\overline{)776}$

Standards-Based Math • 3-4 © 2004 Creative Teaching Press

One-Digit Division—Remainders

NUMBER AND OPERATIONS

1 $2\overline{)5}$ $3\overline{)5}$ $5\overline{)8}$ $4\overline{)9}$

2 $2\overline{)9}$ $3\overline{)8}$ $3\overline{)4}$ $5\overline{)7}$

3 $4\overline{)7}$ $2\overline{)7}$ $5\overline{)6}$ $4\overline{)5}$

4 $7\overline{)9}$ $8\overline{)9}$ $6\overline{)8}$ $7\overline{)8}$

5 $6\overline{)9}$ $5\overline{)9}$ $4\overline{)6}$ $3\overline{)7}$

Rise and Shine

NUMBER AND OPERATIONS

If you live in West Quoddy Head, Maine, you are among the first Americans to see the sun rise every day. Match the correct houses and suns.

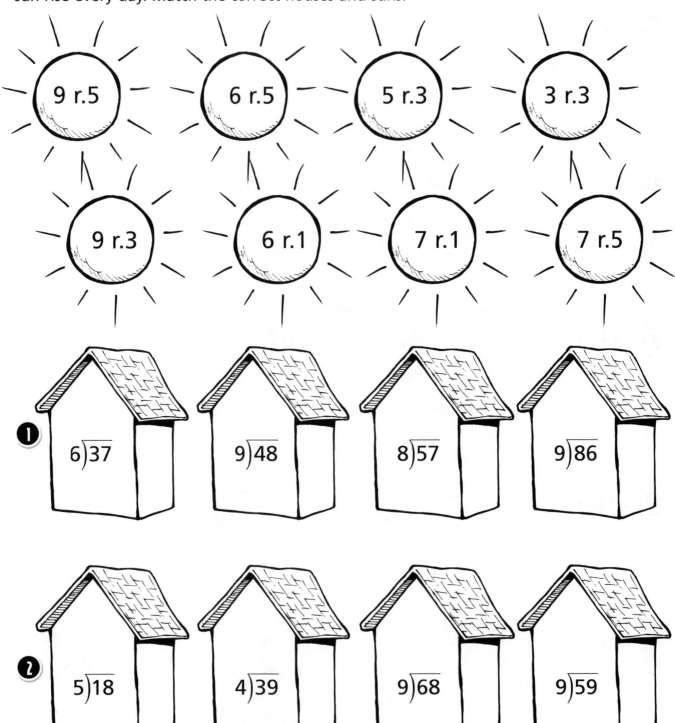

Name _____   Date _____

Three-Digit Division—Remainders

NUMBER AND OPERATIONS

1 2)231   5)632   4)253   3)626

2 4)391   7)284   3)845   6)403

3 3)727   2)305   9)431   5)902

4 8)983   4)306   6)442   5)839

5 3)773   8)451   4)339   7)950

Apple Seed Solutions

NUMBER AND OPERATIONS

Johnny Appleseed planted apple seeds in Ohio and other parts of midwestern America. Solve the problems to learn his real name. Find the letter that matches each answer and write it on the line.

$80 \div 8 =$ ____ $150 \div 10 =$ ____ $80 \div 10 =$ ____ $140 \div 10 =$ ____

$30 \div 10 =$ ____ $160 \div 20 =$ ____ $100 \div 100 =$ ____ $160 \div 10 =$ ____

$30 \div 10 =$ ____ $50 \div 50 =$ ____ $280 \div 20 =$ ____

1	2	3	4	5	6	7	8	9	10	11	12	13
a	b	c	d	e	f	g	h	i	j	k	l	m

14	15	16	17	18	19	20	21	22	23	24	25	26
n	o	p	q	r	s	t	u	v	w	x	y	z

Standards-Based Math • 3–4 © 2004 Creative Teaching Press

Word Problems 4

NUMBER AND OPERATIONS

1 Jamie will serve each guest two hot dogs. Jamie purchased 56 hot dogs. How many people can she serve?

2 The Corliss family needed to fill their gas tank 3 times on their family vacation. Each time they stopped they filled their tank the same amount. They bought a total of 78 gallons of gas. How many gallons of gas did they buy each time?

3 Kristin is moving on Friday. She has 48 books and only 4 boxes. She wants to put an equal number of books in each box. How many books will she pack in each box?

4 Vince is saving money to buy a new bike. He needs $168. How much money does he need to save each day if he wants to buy the bike in a week?

5 Nailah collected 130 shells on her trip. She wants to give the same number of shells to each of her five cousins. There are two cousins in one family, and three cousins in the other. How many shells will the family with three children get?

Adding Fractions with Like Denominators

NUMBER AND OPERATIONS

1 $\dfrac{2}{4} + \dfrac{1}{4}$ $\dfrac{1}{3} + \dfrac{1}{3}$ $\dfrac{1}{5} + \dfrac{2}{5}$

2 $\dfrac{2}{6} + \dfrac{2}{6}$ $\dfrac{3}{6} + \dfrac{2}{6}$ $\dfrac{1}{4} + \dfrac{1}{4}$

3 $\dfrac{3}{9} + \dfrac{2}{9}$ $\dfrac{2}{7} + \dfrac{4}{7}$ $\dfrac{2}{5} + \dfrac{2}{5}$

4 $\dfrac{2}{10} + \dfrac{5}{10}$ $\dfrac{4}{8} + \dfrac{1}{8}$ $\dfrac{3}{6} + \dfrac{1}{6}$

5 $\dfrac{4}{6} + \dfrac{2}{6}$ $\dfrac{1}{4} + \dfrac{1}{4}$ $\dfrac{3}{7} + \dfrac{3}{7}$

Standards-Based Math • 3–4 © 2004 Creative Teaching Press

Subtracting Fractions with Like Denominators

NUMBER AND OPERATIONS

1 $\dfrac{3}{4} - \dfrac{1}{4}$ $\qquad\qquad$ $\dfrac{4}{5} - \dfrac{2}{5}$ $\qquad\qquad$ $\dfrac{5}{6} - \dfrac{2}{6}$

2 $\dfrac{3}{5} - \dfrac{2}{5}$ $\qquad\qquad$ $\dfrac{4}{7} - \dfrac{2}{7}$ $\qquad\qquad$ $\dfrac{4}{6} - \dfrac{2}{6}$

3 $\dfrac{6}{9} - \dfrac{2}{9}$ $\qquad\qquad$ $\dfrac{7}{8} - \dfrac{5}{8}$ $\qquad\qquad$ $\dfrac{5}{7} - \dfrac{2}{7}$

4 $\dfrac{7}{6} - \dfrac{2}{6}$ $\qquad\qquad$ $\dfrac{3}{9} - \dfrac{2}{9}$ $\qquad\qquad$ $\dfrac{7}{8} - \dfrac{5}{8}$

5 $\dfrac{4}{5} - \dfrac{1}{5}$ $\qquad\qquad$ $\dfrac{8}{10} - \dfrac{5}{10}$ $\qquad\qquad$ $\dfrac{4}{7} - \dfrac{3}{7}$

Adding Fractions with Unlike Denominators

NUMBER AND OPERATIONS

1 $\dfrac{1}{3} + \dfrac{1}{6}$ $\qquad\qquad$ $\dfrac{1}{2} + \dfrac{1}{6}$ $\qquad\qquad$ $\dfrac{1}{2} + \dfrac{1}{4}$

2 $\dfrac{2}{6} + \dfrac{1}{12}$ $\qquad\qquad$ $\dfrac{1}{2} + \dfrac{1}{8}$ $\qquad\qquad$ $\dfrac{2}{9} + \dfrac{2}{3}$

3 $\dfrac{1}{3} + \dfrac{1}{2}$ $\qquad\qquad$ $\dfrac{1}{4} + \dfrac{2}{3}$ $\qquad\qquad$ $\dfrac{1}{2} + \dfrac{1}{5}$

4 $\dfrac{1}{2} + \dfrac{2}{4}$ $\qquad\qquad$ $\dfrac{1}{3} + \dfrac{2}{5}$ $\qquad\qquad$ $\dfrac{1}{2} + \dfrac{2}{7}$

Standards-Based Math • 3–4 © 2004 Creative Teaching Press

Subtracting Fractions with Unlike Denominators

NUMBER AND OPERATIONS

1 $\dfrac{6}{8} - \dfrac{2}{4}$ $\dfrac{3}{5} - \dfrac{1}{10}$ $\dfrac{5}{6} - \dfrac{1}{3}$

2 $\dfrac{3}{4} - \dfrac{1}{8}$ $\dfrac{1}{2} - \dfrac{1}{3}$ $\dfrac{2}{3} - \dfrac{1}{4}$

3 $\dfrac{4}{5} - \dfrac{1}{2}$ $\dfrac{4}{5} - \dfrac{2}{10}$ $\dfrac{5}{6} - \dfrac{2}{3}$

4 $\dfrac{3}{4} - \dfrac{1}{3}$ $\dfrac{1}{2} - \dfrac{2}{5}$ $\dfrac{2}{3} - \dfrac{1}{5}$

Reducing Fractions

Number and Operations

1 $\dfrac{2}{4} =$ $\dfrac{3}{6} =$ $\dfrac{2}{6} =$ $\dfrac{4}{12} =$

2 $\dfrac{7}{21} =$ $\dfrac{5}{15} =$ $\dfrac{4}{6} =$ $\dfrac{3}{9} =$

3 $\dfrac{3}{12} =$ $\dfrac{6}{8} =$ $\dfrac{8}{12} =$ $\dfrac{6}{16} =$

4 $\dfrac{4}{14} =$ $\dfrac{6}{18} =$ $\dfrac{8}{18} =$ $\dfrac{6}{15} =$

5 $\dfrac{5}{25} =$ $\dfrac{7}{14} =$ $\dfrac{8}{24} =$ $\dfrac{9}{18} =$

6 $\dfrac{4}{28} =$ $\dfrac{12}{16} =$ $\dfrac{6}{32} =$ $\dfrac{10}{45} =$

Standards-Based Math • 3–4 © 2004 Creative Teaching Press

Word Problems 5

Number and Operations

1 Jamie invited 28 people to her birthday party, but only half of the people can make it. How many people should Jamie expect at her party?

2 Alexander has four sisters. Ann has one brother. If Ann has one-half as many sisters as Alexander, how many sisters does Ann have?

3 Morgan walks 9 blocks to get to school. She meets Isabella one-third of the way there. How many blocks does Morgan walk before she meets Isabella?

4 Grandma's wonder cake recipe calls for $\frac{1}{2}$ cup of sugar and $\frac{2}{3}$ cup of flour. How many cups of dry ingredients does Grandma need for her cake?

5 Jackie filled her water bottle $\frac{3}{4}$ of the way with water. She drank half of that at halftime. How much water was left for Jackie at the end of the game?

Repeated Patterns

Algebra

What comes next?

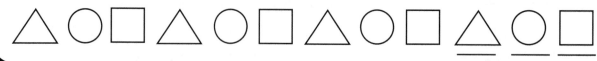

Draw the next three images in each pattern.

1  ___ ___ ___

2 A B D C A B D C A ___ ___ ___

3  ___ ___ ___

4  ___ ___ ___

5 + ÷ × = + ÷ × = + ÷ ___ ___ ___

6  ___ ___ ___

Standards-Based Math • 3–4 © 2004 Creative Teaching Press

Number Patterns

ALGEBRA

4, 8, 12, 16, 20, 24

pattern: +4

Continue the pattern.

1 2, 4, 6, 8, _____, _____, _____, _____

2 100, 95, 90, 85, 80, _____, _____, _____, _____

3 1, 3, 6, 9, 12, _____, _____, _____, _____

4 2, 7, 12, 17, 22, _____, _____, _____, _____

5 1, 2, 4, 8, 16, _____, _____, _____, _____

6 50, 48, 46, 44, _____, _____, _____, _____

7 10, 20, 30, 40, _____, _____, _____, _____

8 200, 175, 150, 125, _____, _____, _____, _____

9 1, 5, 6, 10, 11, 15, 16, _____, _____, _____, _____

10 2, 4, 5, 10, 11, 22, 23, _____, _____, _____, _____

Name _____   Date _____

Number Chart Patterns

Algebra

Complete the number chart.

1	2	3	4	5	6	7	8	9	10
11									
21	22		24						
31								39	
41			45						
51									
61						67			
71		73							80
81	82								
91									100

Complete the following steps. Notice the color patterns you create.

1 Color the multiples of 5 **green**.

2 Color the even numbers **blue**.

3 Color the prime numbers **red**.

4 Draw a **purple X** over the numbers that are divisible by 4.

Standards-Based Math • 3–4 © 2004 Creative Teaching Press

Positive and Negative Integers

Algebra

> Positive numbers are greater than 0.
> Negative numbers are less than 0.
> Positive and negative numbers can be shown on a number line.
>
> Less than 0   Greater than 0
>
>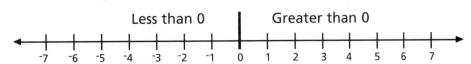

Use the number line to compare each set of numbers. Use the <, >, or = sign to complete each comparison.

1 4 ◯ 5

2 ⁻12 ◯ ⁻10

3 1 ◯ ⁻1

4 ⁻5 ◯ ⁻8

5 6 ◯ 7

6 0 ◯ 8

7 ⁻3 ◯ 3

8 ⁻7 ◯ 7

9 ⁻12 ◯ ⁻1

10 ⁻9 ◯ ⁻3

11 9 ◯ ⁻11

12 8 ◯ ⁻8

13 5 ◯ 5

14 ⁻7 ◯ 2

15 ⁻6 ◯ ⁻4

Order Properties

Algebra

When you add or multiply numbers, you can change their order without changing the answer.

$4 \times 3 = 3 \times 4$ $<$ $4 \times 3 = 12$
$3 \times 4 = 12$

$2 + 4 = 6$ $>$ $2 + 4 = 4 + 2$
$4 + 2 = 6$

When you subtract and divide numbers, you cannot change the order of the numbers without changing the answer.

$8 - 5 \neq 5 - 8$ $<$ $8 - 5 = 3$
$5 - 8 = {}^-3$

$6 \div 3 = 2$ $>$ $6 \div 3 \neq 3 \div 6$
$3 \div 6 = \frac{1}{2}$

Use the order properties to decide if each statement is true or false.
Write **T** it the statement is **true**. Write **F** if it is **false.**

___ **1** $3 + 1 = 1 + 3$

___ **6** $4 + 12 = 12 + 4$

___ **2** $7 - 5 = 5 - 7$

___ **7** $18 \div 6 = 6 \div 18$

___ **3** $10 + 12 = 12 + 10$

___ **8** $3 \times 9 = 9 \times 3$

___ **4** $2 \times 24 = 24 \times 2$

___ **9** $4 - 18 = 18 - 4$

___ **5** $100 \div 2 = 2 \div 100$

___ **10** $1 - 7 = 7 - 1$

Standards-Based Math • 3–4 © 2004 Creative Teaching Press

Equality Properties

Algebra

An equation is like a balance scale. Both sides must be equal to make the scale balance.

$2 = 2$

$2 + 1 \neq 2$ If you change only one side of the equation, the scale is not balanced and the equation is not true.

$2 + 1 = 2 + 1$ Making the same change on both sides of the equation will make it equal and true.

Fill in the number that makes each equation true.

1 $5 + 3 = \underline{\hspace{1cm}} + 3$

2 $6 \div 3 = 6 \div \underline{\hspace{1cm}}$

3 $\underline{\hspace{1cm}} - 2 = 10 - 2$

4 $7 + 2 = 2 + \underline{\hspace{1cm}}$

5 $\underline{\hspace{1cm}} \times 8 = 8 \times 4$

6 $9 - \underline{\hspace{1cm}} = 9 - 1$

7 $24 \div \underline{\hspace{1cm}} = 24 \div 8$

8 $11 + 7 = \underline{\hspace{1cm}} + 11$

9 $56 \times 4 = \underline{\hspace{1cm}} \times 4$

10 $100 \div 25 = \underline{\hspace{1cm}} \div 25$

Missing Addends and Subtrahends

Algebra

Solve each equation. Then write the letter on the line for each answer. The letters will spell out the answer to the riddle.

K ① $10 = 4 +$ ____

R ② $5 +$ ____ $= 20$

E ③ $50 -$ ____ $= 42$

S ④ $36 = 12 +$ ____

C ⑤ $15 +$ ____ $= 46$

C ⑥ $18 -$ ____ $= 8$

A ⑦ ____ $+ 9 = 23$

O ⑧ $48 = 36 +$ ____

A ⑨ $22 -$ ____ $= 13$

D ⑩ $15 +$ ____ $= 38$

F ⑪ $19 +$ ____ $= 26$

D ⑫ ____ $- 30 = 11$

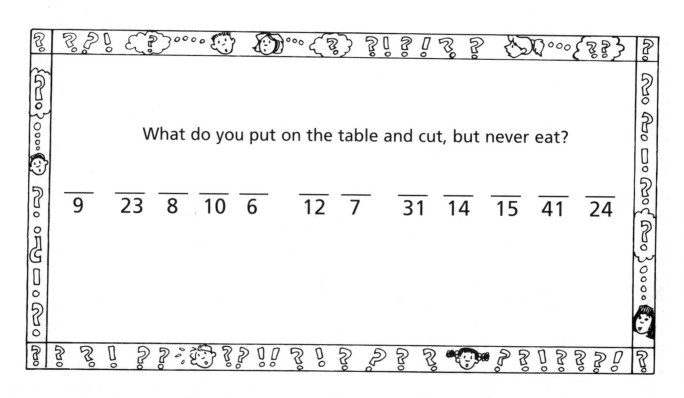

What do you put on the table and cut, but never eat?

___ ___ ___ ___ ___   ___ ___   ___ ___ ___ ___ ___
9   23   8   10   6   12   7   31   14   15   41   24

Standards-Based Math • 3–4 © 2004 Creative Teaching Press

Missing Factors and Divisors

Algebra

Solve each equation. Then write the letter on the line for each answer. The letters will spell out the answer to the riddle.

A **1** $20 = 5 \times$ _____

A **2** $8 = 24 \div$ _____

T **3** $44 =$ _____ $\times 4$

B **4** $12 \times$ _____ $= 60$

U **5** $100 \div$ _____ $= 5$

K **6** _____ $\times 7 = 63$

R **7** $6 \times$ _____ $= 72$

G **8** _____ $\times 8 = 56$

G **9** $64 \div$ _____ $= 8$

C **10** $54 \div$ _____ $= 9$

E **11** $3 \times$ _____ $= 39$

R **12** $36 \div$ _____ $= 2$

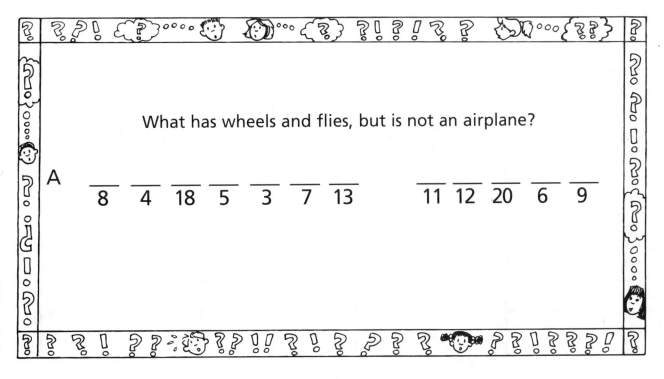

What has wheels and flies, but is not an airplane?

A $\dfrac{}{8}$ $\dfrac{}{4}$ $\dfrac{}{18}$ $\dfrac{}{5}$ $\dfrac{}{3}$ $\dfrac{}{7}$ $\dfrac{}{13}$ $\dfrac{}{11}$ $\dfrac{}{12}$ $\dfrac{}{20}$ $\dfrac{}{6}$ $\dfrac{}{9}$

Function Tables

A̲LGEBRA

Complete each function table.

① Rule is + 4

In	Out
2	
5	
6	
9	
12	
8	

③ Rule is - 4

In	Out
10	
7	
13	
21	
17	
9	

⑤ Rule is + 11

In	Out
1	
7	
21	
12	
9	
4	

② Rule is × 7

In	Out
2	
4	
1	
0	
11	
6	

④ Rule is ÷ 2

In	Out
14	
28	
10	
16	
8	
6	

⑥ Rule is × 8

In	Out
2	
8	
3	
5	
7	
9	

Standards-Based Math • 3–4 © 2004 Creative Teaching Press

What's the Rule?

Algebra

Write the rule for each function table.

1 Rule is ____

In	Out
7	13
4	10
9	15
2	8
5	11
6	12

3 Rule is ____

In	Out
20	5
36	9
4	1
12	3
28	7
16	4

5 Rule is ____

In	Out
18	15
21	18
6	3
12	9
8	5
13	10

2 Rule is ____

In	Out
2	10
4	20
7	35
9	45
3	15
8	40

4 Rule is ____

In	Out
6	42
9	63
11	77
2	14
5	35
7	49

6 Rule is ____

In	Out
24	12
48	24
18	9
14	7
28	14
8	4

Name _____ Date _____

Find the Rule

ALGEBRA

Write the rule for each function table and complete the table.

1 Rule is ___

In	Out
4	36
9	81
3	
	54
	9
7	63

2 Rule is ___

In	Out
11	20
2	11
	10
5	
8	17
	18

3 Rule is ___

In	Out
18	22
7	11
	9
	24
11	
3	7

4 Rule is ___

In	Out
10	8
16	14
	11
9	
20	
4	2

5 Rule is ___

In	Out
24	4
30	5
60	
	6
18	3
6	

6 Rule is ___

In	Out
18	9
26	17
9	
	11
	3
10	1

Standards-Based Math • 3–4 © 2004 Creative Teaching Press

Name _____ Date _____

Writing Expressions

ALGEBRA

You can use letters to express unknowns, or variables, in an equation.

Example: There are 2 more apples than bananas. Write an expression to show how many apples there are.

number of apples = 2 + b   b = number of bananas

Write an expression for each story problem.

1 Each student has 3 more blue pens than red pens. Write an expression to show how many red pens each student has.
r = number of red pens   _____

2 Mike had 10 crackers for a snack. He ate some of the crackers. Write an expression to show how many crackers Mike has left.
n = number of crackers Mike ate   _____

3 Erin has 4 times as many dolls as cars. Write an expression to show the number of dolls Erin has.
c = number of cars   _____

4 Three people share a pizza equally. Write an expression to show how many pieces of pizza each person gets.
p = total number of pieces of pizza   _____

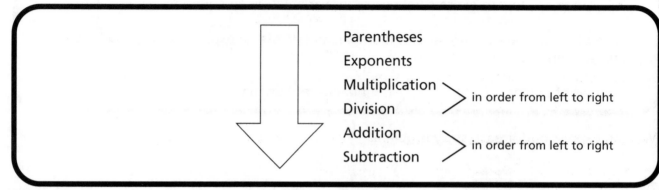

Order of Operations

Algebra

Solve the problems according to the order of operations listed in the box.

Parentheses

Exponents

Multiplication
Division
> in order from left to right

Addition
Subtraction
> in order from left to right

1 $4 + (12 \div 2) \times 3 =$

2 $8 + 2 \times 4 + 1 =$

3 $10 \div 2 - 4 + 1 =$

4 $\frac{1}{2}(3 + 3) \times 5 =$

5 $\frac{1}{3}(30 \div 2) =$

6 $12 - 2 + 4 \times 2 =$

7 $8 + (7 - 2) \times 2 =$

8 $2(3 + 1 + 2) =$

9 $(10 - 4) \div 2 - 1 =$

10 $3(5 - 1) \times 2 - 4 =$

11 $4 \div 2 + (2 \times 5) =$

12 $2 \times 3 \times 3^2 =$

13 $(3 + 2)^2 =$

14 $2(3 + 1)^2 - 8 =$

Standards-Based Math • 3–4 © 2004 Creative Teaching Press

Writing Equations with Variables

Algebra

An equation is a mathematical sentence that shows two equal expressions.

equals sign

$\underbrace{5 + 4}_{\text{expression}} \overset{|}{=} \underbrace{9}_{\text{expression}}$

Sometimes an equation contains a variable.

variable

$5 + \overset{|}{n} = 9$

Write an equation to represent each story problem.

1 Mary has 3 oranges. She buys more to make a total of 7 oranges. How many does she buy?

equation: _____   n = number of oranges she buys

2 Matt has 39 baseball cards. He gave away some to his friends. Now he has 28 cards. How many did he give away?

equation: _____   c = number of cards he gave away

3 Cara sold 4 times as many candy bars as Sarah. Cara sold 28 candy bars. How many did Sarah sell?

equation: _____   b = number of candy bars Sarah sold

4 Max brought marbles to school. He divided them equally between his 8 friends. If each friend received 8 marbles, how many marbles did Max bring?

equation: _____   m = number of marbles

Writing and Solving Equations

ALGEBRA

Write an equation for each story problem. Then solve it.

1 Braden Elementary School collected cans. The third grade collected 125 cans. The fourth grade collected more. Both classes collected a total of 300 cans. How many did the fourth grade collect?
c = the number of cans collected by fourth graders

equation: _____   c = _____

2 Kelsey jumped 3 times farther than Mike. Kelsey's record jump was 36 inches. How far did Mike jump?
m = how far Mike jumped

equation: _____   m = _____

3 The custodian needed to set up 49 chairs. Each row has 7 chairs. How many rows of chairs must the custodian set up?
r = number of rows of chairs

equation: _____   r = _____

4 The library has 251 new books. Students checked out some of the books. The librarian counted 114 books left. How many books were checked out?
b = number of books checked out

equation: _____   b = _____

5 A school bus carrying 27 students came to a stop. No new students got on the bus, but some did get off. The bus now has 12 students on board. How many students got off the bus?
s = number of students who got off

equation: _____   s = _____

Standards-Based Math • 3–4 © 2004 Creative Teaching Press

Solving Equations 1

Algebra

Use what you know about addition and subtraction to solve the equations.

1 $20 - n = 15$
n =

2 $n + 8 = 27$
n =

3 $n + 3 = 5 + 1$
n =

4 $4 + n = 8 + 2$
n =

5 $12 - n = 10$
n =

6 $n + 8 = 20 - 4$
n =

7 $n + 9 = 30$
n =

8 $18 - 7 = n$
n =

9 $31 + 13 = n$
n =

10 $9 + 4 = 10 + n$
n =

11 $5 + n = 28$
n =

12 $12 + 24 = n$
n =

13 $7 + n = 13 + 1$
n =

14 $29 - n = 20 + 5$
n =

15 $17 - n = 8$
n =

16 $46 + 11 = n$
n =

17 $19 - 9 = n + 5$
n =

18 $21 - 1 = n + 7$
n =

19 $n = 31 + 14$
n =

20 $7 + n = 28$
n =

Name _____ Date _____

Solving Equations 2

Algebra

Solve the equations.

1 $4 \times n = 28$
n =

2 $84 \div n = 21$
n =

3 $3 \times 5 = n$
n =

4 $9 \times n = 54$
n =

5 $n \div 6 = 36$
n =

6 $n + 7 = 2 \times 4$
n =

7 $24 \div n = 5 + 1$
n =

8 $n \times 3 = 18 + 3$
n =

9 $14 \times 4 = n$
n =

10 $15 \times n = 90$
n =

11 $27 - n = 5 \times 4$
n =

12 $n \times 7 = 40 + 9$
n =

13 $8 \times n = 50 + 6$
n =

14 $18 \div n = 18 - 9$
n =

15 $21 \div 3 = 14 - n$
n =

16 $11 \times 11 = 100 + n$
n =

17 $9 \times n = 44 + 55$
n =

18 $48 \div n = 3 \times 4$
n =

19 $212 - 152 = 12 \times n$
n =

20 $(4 \times n) + 3 = 5 \times 3$
n =

Formulas

A<small>LGEBRA</small>

> A formula is a rule that is written as an equation.
>
> P = 2l + 2w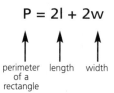
>
> perimeter length width
> of a
> rectangle
>
> A = l × w
>
> area length width

Use the formula for area and perimeter of a rectangle to solve the following problems.

1 The school swimming pool is 55 yards long and 23 yards wide. What is its perimeter?

equation: _____   p = _____

2 The classroom is 22 feet long and 15 feet wide. What is the area of the classroom?

equation: _____   a = _____

3 Melissa's garden is 6 feet long and 4 feet wide. She wants to know how much fencing to buy in order to build a fence around the garden. What is the garden's perimeter?

equation: _____   p = _____

4 Mr. Allen is putting a carpet in his living room. The room is 16 feet long and 15 feet wide. Find the area of the room.

equation: _____   a = _____

Find That Item

Algebra

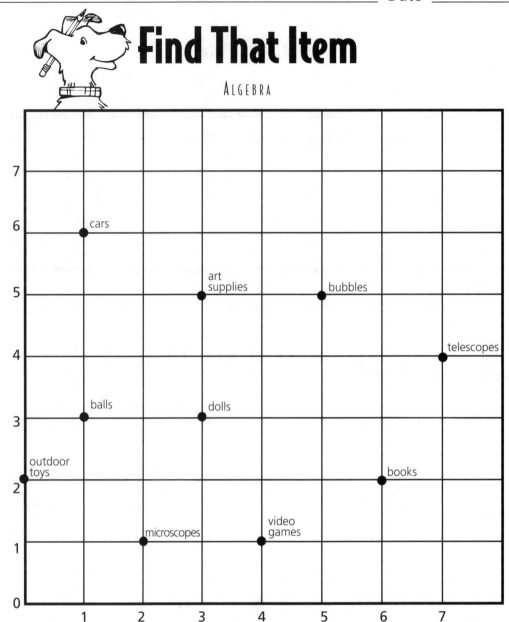

Use the grid of the toy store stock room to locate each item. Record the ordered pair of numbers for each item.

1 outdoor toys _____

2 microscopes _____

3 bubbles _____

4 cars _____

5 balls _____

6 video games _____

7 telescopes _____

8 art supplies _____

9 dolls _____

10 books _____

Standards-Based Math • 3–4 © 2004 Creative Teaching Press

Mystery Picture

ALGEBRA

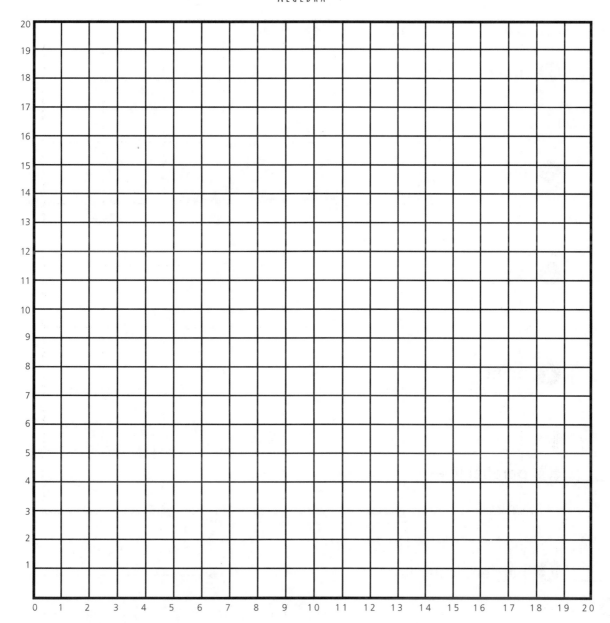

Plot the ordered pairs in the order they are listed. Connect the dots as you find each one. Start a new figure at each arrow.

→ (6, 12) (8, 14) (10, 16) (12, 14) (12, 16) (13, 16) (13, 13)
(14, 12) (14, 9) (14, 5) (11, 5) (11, 8) (9, 8) (9, 5) (6, 5) (6, 9) (6,12)

→ (8, 11) (9, 11) (9, 10) (8, 10) (8, 11)

→ (11, 10) (12, 10) (12, 11) (11, 11) (11, 10)

→ (6, 12) (10, 12) (14, 12)

What picture did you create? _____

Standards-Based Math • 3–4 © 2004 Creative Teaching Press

Language of Lines

Geometry

Match each vocabulary term with the correct example.

_____ **1** line segment   a.

_____ **2** line   b.   •B

_____ **3** ray   c.

_____ **4** point   d.

_____ **5** parallel lines   e.

_____ **6** perpendicular lines   f. •————————•

_____ **7** intersecting lines   g.

Standards-Based Math • 3–4 © 2004 Creative Teaching Press

Connect the Line Segments

GEOMETRY

Draw each line segment using a straightedge.

•M   •J

•A

•N   •K

•L   •C   •G
        •B

        •D
                        •H
        •E
•Q
                    •F
•P
                                    •I

        •S

•O

                        •R

•T

AB	CD	DR	HI
BC	DF	FO	ST
AC	CG	EO	PQ
BE	DG	EL	MN
EF	FR	BL	KJ

What picture did you create? _____

Name _____   Date _____

Identify the Angles

GEOMETRY

obtuse angle	=	angle that measures greater than 90°, but less than 180°
acute angle	=	angle that measures less than 90°
right angle	=	angle that measures exactly 90°
straight angle	=	angle that measures exactly 180°

Identify each angle by writing **obtuse, acute, right,** or **straight**.

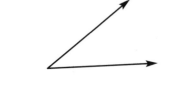

1 _____

5 _____

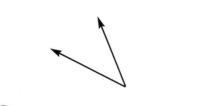

2 _____

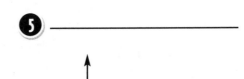

6 _____

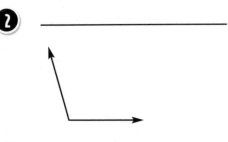

3 _____

7 _____

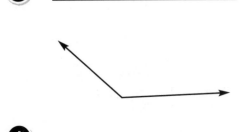

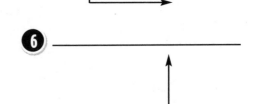

4 _____

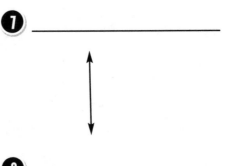

8 _____

Standards-Based Math • 3–4 © 2004 Creative Teaching Press

Identify the Shapes

Geometry

Color each shape as follows:

octagon: purple
trapezoid: brown
oval: green
triangle: blue
hexagon: orange
pentagon: yellow
circle: pink

Polygon or Not?

GEOMETRY

A **polygon** is a closed figure with sides that are all line segments.

Polygons	Not Polygons
_____	_____
_____	_____
_____	_____
_____	_____
_____	_____
_____	_____
_____	_____

Sort each shape into the correct category. Write the letter of the shape in the correct column.

A

B

C

D

E

F

G

H

I

J

K

L

Name _____ Date _____

Draw That Shape

GEOMETRY

Draw an example of each figure.

Trapezoid	Square	Hexagon
Parallelogram	Rectangle	Circle
Rhombus	Quadrilateral	Oval
Triangle	Pentagon	Polygon

List all of the figure names from above that can be used to describe a **square**.

_____   _____

_____   _____

_____

Standards-Based Math • 3–4 © 2004 Creative Teaching Press

Naming Triangles

Geometry

Triangles can be named based on the length of their sides.

Equilateral triangle:
all sides are equal

Isosceles triangle:
two sides are the same
length

Scalene triangle:
each side is a different
length

Label each triangle based on the length of its sides.

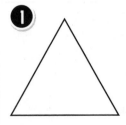 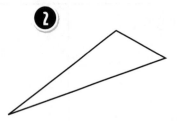 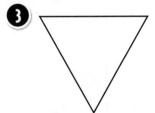 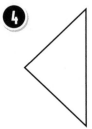

_____   _____   _____   _____

Triangles can also be named by their angle measurements.
Right triangle: one angle measures exactly 90°
Acute triangle: all angles measure less than 90°
Obtuse triangle: one angle measures more than 90°

Label each triangle based on its angle measurements.

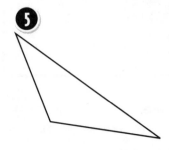 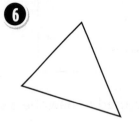 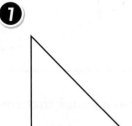 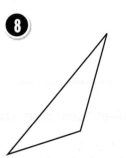

_____   _____   _____   _____

Standards-Based Math • 3–4 © 2004 Creative Teaching Press

Name _____ Date _____

Label the Circle

Geometry

Follow the directions to correctly draw the parts of the circle.

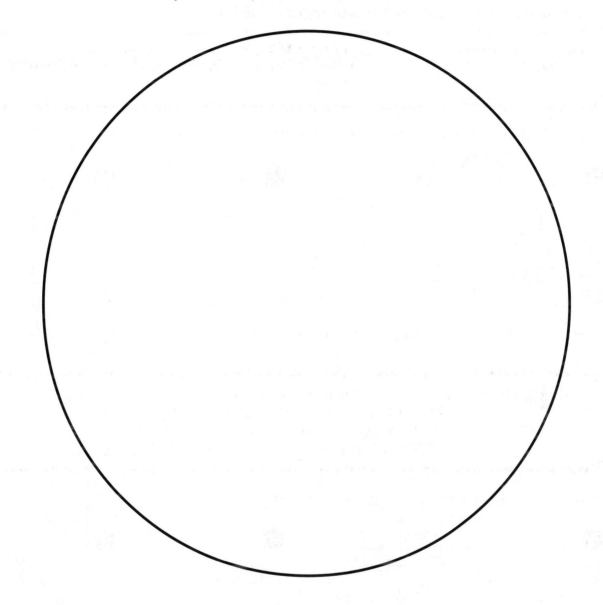

1 Use **green** to draw and label the center of the circle.

2 Use **orange** to draw and label the diameter of the circle.

3 Use **black** to draw and label the radius of the circle.

4 Use **purple** to draw and label a chord on the circle.

Congruent or Not?

GEOMETRY

Congruent figures have the same shape and size.

Write **yes** after each pair of figures if they are congruent. Write **no** if they are not congruent.

1 _____

2 _____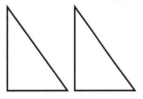

3 _____

4 _____

5 _____

6 _____

7 _____

8 _____

9 _____

10 _____

Standards-Based Math • 3–4 © 2004 Creative Teaching Press

Slides, Flips, and Turns

Geometry

Slides
When you slide a figure, it does not change in any way except that it is now in a different place.

Flips
When you flip a figure across a line, it looks like a reflection of itself.

Turns
When you turn a figure, you spin it around a point.

Label each drawing, as a **slide**, **flip**, or **turn** to describe how it was moved.

1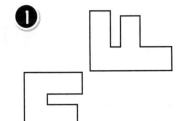

_____

2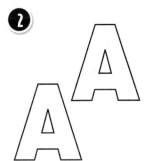

_____

3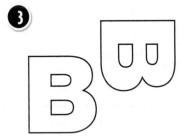

_____

4

_____

5

_____

6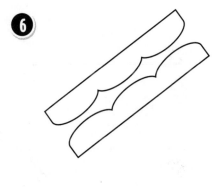

_____

Name _____ Date _____

Shape Puzzler

GEOMETRY

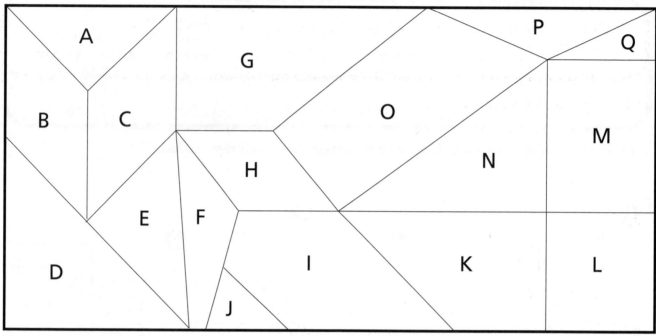

Write the letter to tell where each piece fits. You can slide, turn, and flip the pieces.

 ❶
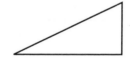

_____

❷
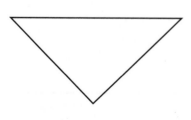

_____

❸
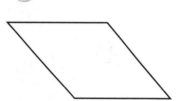

_____

❹
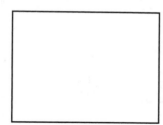

_____

❺

_____

❻
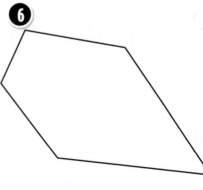

_____

Standards-Based Math • 3–4 © 2004 Creative Teaching Press

Name _____ Date _____

Make Similar Figures

Geometry

Similar figures have the same shape. They may be the same size or they may have different sizes.

Draw a similar figure for each figure.

1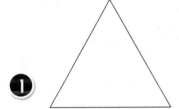

6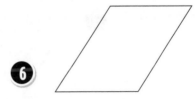

2

7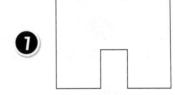

3

8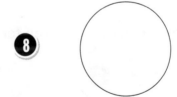

4

9

5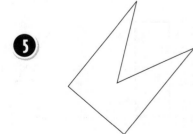

10

Name _____ Date _____

Lines of Symmetry

Geometry

This figure is symmetrical.

The dotted line is a line of symmetry.

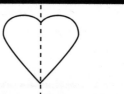

A. Write **yes** if the dotted line is a line of symmetry. Write **no** if it is not a line of symmetry.

❶ _____

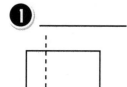

❷ _____

❸ _____

B. Each figure has two lines of symmetry. One is drawn for you. Draw the second line of symmetry.

❹

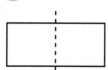

❺

❻

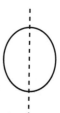

C. Draw as many lines of symmetry as you can for each figure.

❼

❽

❾

D. Draw the other half of each figure.

❿

⓫

⓬

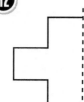

Standards-Based Math • 3–4 © 2004 Creative Teaching Press

Name _____ Date _____

How Many Lines?

Geometry

Write whether each figure has 0, 1, 2, 3, 4, or more than four lines of symmetry.

1  _____

2  _____

3  _____

4 _____

5  _____

6  _____

7  _____

8  _____

9 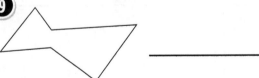 _____

10 _____

Parts of Solid Figures

Geometry

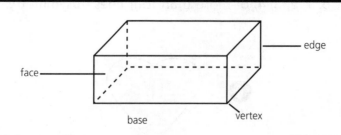

A **face** is the solid flat surface of a figure.
An **edge** is where two faces meet.
A **vertex** is the point where two edges meet. The plural of "vertex" is "vertices."
A **solid** figure rests on its base.

Complete the chart.

Solid Figure	Shape of Base	Number of Faces	Number of Vertices	Number of Edges

What Am I?

GEOMETRY

Label each everyday item by its solid figure. Use **cylinder, cone, sphere, pyramid,** or **cube.**

1  _____

2 _____

3 _____

4 _____

5 _____

6 _____

7 _____

8 _____

9  _____

10 _____

Identifying Bases

Geometry

Draw a line between each figure and the shape of its base.

 1

 2

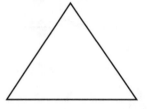

 3

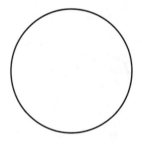

 4

 **5**

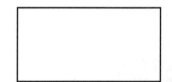

 6

Name _____ Date _____

Draw That Figure

GEOMETRY

Draw an example of each solid figure.

Prism	Sphere
Cylinder	Cone
Cube	Pyramid

Write two names that describe this figure.

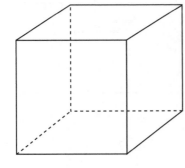

_____

_____

Name _____ Date _____

Solid Figure Riddle

GEOMETRY

Match each object below to the correct solid figure. Write the solid figure's letter on the line above the item to reveal the answer to the riddle.

S

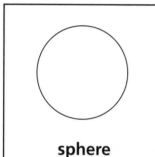

sphere

O

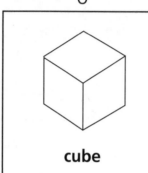

cube

P

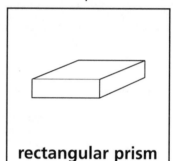

rectangular prism

F

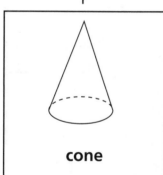

cone

E

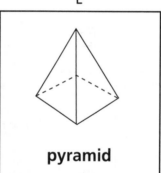

pyramid

T

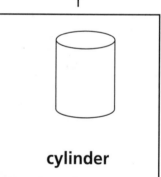

cylinder

Riddle: The more you take, the more you leave behind. What are they?

___ ___ ___ ___ ___ ___ ___ ___ ___

Standards-Based Math • 3–4 © 2004 Creative Teaching Press

Similar or Congruent?

GEOMETRY

Label each pair of figures as **similar** or **congruent**.

1 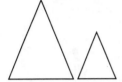   _____

2   _____

3    _____

4 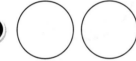   _____

5   _____

6    _____

7   _____

8 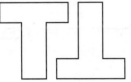   _____

9   _____

10   _____

Standards-Based Math • 3–4 © 2004 Creative Teaching Press

What Time Is It?

MEASUREMENT

Write the time shown on each clock.

1

_____

4

_____

7

_____

2

_____

5

_____

8

_____

3

_____

6

_____

9

_____

Standards-Based Math • 3–4 © 2004 Creative Teaching Press

Name _____ Date _____

Figure the Time

MEASUREMENT

Use the clock face to help you answer the word problems. Write your answer on the line after each problem.

1 Sam's bus arrives at 8:45 a.m. He needs to wake up 30 minutes before the bus comes. What time should he get up? _____

2 School begins at 9:00 a.m. Students spend 6 hours at school. What time does school end? _____

3 Lunch is 45 minutes long. If lunch ends at 1:15 p.m., what time does it begin? _____

4 Students spend 55 minutes in math class and 55 minutes in reading class. How long do they spend in reading and math together? _____

5 Sam gets on the bus at 3:15 p.m. If his bus trip lasts 24 minutes, what time will he get home? _____

6 Sam has baseball practice at 4:30 p.m. Practice lasts 45 minutes. If dinner is served at 5:30 p.m. How long will Sam have to get home from practice? _____

Name _____ Date _____

Calendar Days

MEASUREMENT

Use the calendar to answer the questions.

November 2005						
S	M	T	W	T	F	S
		1	2	3	4	5
6	7	8	9	10	11	12
13	14	15	16	17	18	19
20	21	22	23	24	25	26
27	28	29	30			

1 On November 1, Sandy made plans with Bill to meet for lunch in 2 weeks. On what date are they planning to meet?

_____

2 On November 11, Sandy remembered that she missed a dentist appointment 6 days earlier. What was the date of her missed appointment?

_____

5 Thanksgiving is always the fourth Thursday in November. What is the date of Thanksgiving in 2005?

_____

3 Bill checks a book out of the school library on November 4. It is due back in 3 weeks. What is the date that the book is due?

_____

6 Sandy is going to a concert on the third Monday in November. On what date is she going to the concert?

_____

4 On November 21, Bill finds a library book that was due on November 6. How many days overdue is this book?

_____

7 On November 3, Bill made an appointment to meet his accountant in 2 weeks. What is the date of his appointment?

_____

86

Standards-Based Math • 3–4 © 2004 Creative Teaching Press

Name _____   Date _____

Nonstandard Units of Measure

MEASUREMENT

Long ago people measured lengths with units such as cubits, palms, and spans.

cubit

palm

span

Measure each item using cubits, palms, and spans. Record each measurement, and then compare your answers with a classmate's answers.

1 a desk

_____ cubits

_____ palms

_____ spans

2 a window

_____ cubits

_____ palms

_____ spans

3 a textbook

_____ cubits

_____ palms

_____ spans

4 a door

_____ cubits

_____ palms

_____ spans

5 a chalkboard or dry erase board

_____ cubits

_____ palms

_____ spans

6 the height of a chair

_____ cubits

_____ palms

_____ spans

Name _____ Date _____

Customary Measurement

MEASUREMENT

Estimate the length of each line in inches. Then use a ruler to measure each line to the nearest quarter inch.

Object	Estimate	Actual Length
1		
2		
3		
4		
5		
6		

Standards-Based Math • 3–4 © 2004 Creative Teaching Press

Metric Measurement

MEASUREMENT

Estimate the length of each line in centimeters. Then use a metric ruler to measure each line to the nearest half centimeter.

Object	Estimate	Actual Length
1		
2		
3		
4		
5		
6		

Computing Customary Units

MEASUREMENT

Customary Units of Length

inch (in)
foot (ft) = 12 inches
yard (yd) = 3 feet
mile (mi) = 1,760 yards

Use the information from the chart to complete each conversion.

1 24 inches = _____ ft

2 1 mile = _____ ft

3 7 feet = _____ in

4 4 feet = _____ in

5 3 yards = _____ ft

6 2 miles = _____ yds

7 5280 yards = _____ mi

8 60 inches = _____ ft

9 9 yards = _____ ft

10 24 feet = _____ yds

Select the appropriate unit of measurement for each object. Write **inches, feet, yards,** or **miles**.

11 a pencil _____

12 length of room _____

13 football field _____

14 distance between two cities

_____

15 your height _____

16 length of your arm _____

17 distance to an
amusement park _____

18 fabric for
making curtains _____

Computing Metric Units

MEASUREMENT

Metric Units of Length

meter (m)
millimeter (mm)   =   1/1000 of a meter
centimeter (cm)   =   1/100 of a meter
decimeter (dm)   =   1/10 of a meter
kilometer (km)   =   1000 meters

Use the information in the chart to complete each conversion.

1 1 meter = _____ cm

2 40 decimeters = _____ m

3 3 kilometers = _____ m

4 7 meters = _____ mm

5 20 meters = _____ cm

6 2000 millimeters = _____ m

7 4 kilometers = _____ m

8 300 decimeters = _____ m

9 1 decimeter = _____ cm

10 2 centimeters = _____ mm

Order the units from smallest to largest.

decimeter   _____

kilometer   _____

meter   _____

centimeter   _____

millimeter   _____

Perimeter

MEASUREMENT

Perimeter is a measurement of the distance around an object.

Find the perimeter of each object

1
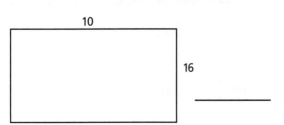

_____

5
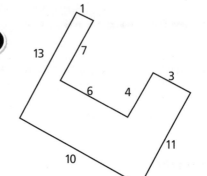

_____

2
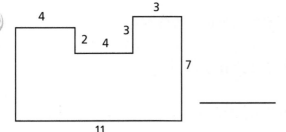

_____

6
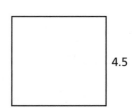

_____

3
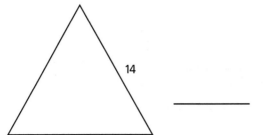

_____

7
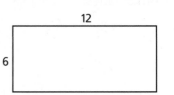

_____

4
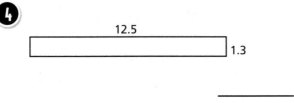

_____

8
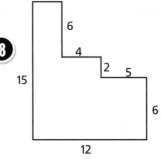

_____

Standards-Based Math • 3-4 © 2004 Creative Teaching Press

Name _____ Date _____

Area 1

Area is the number of square units needed to cover a figure.

 = 3 square units

Find the area by counting the square units.

1

_____

5

_____

9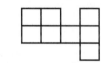

_____

2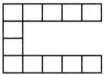

_____

6

_____

10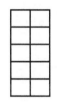

_____

3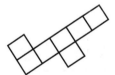

_____

7

_____

11

_____

4

_____

8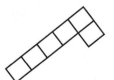

_____

12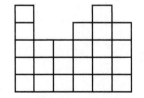

_____

Area 2

MEASUREMENT

Sometimes the area of a figure is made up of half squares as well as whole squares.

1. Count the whole squares: 2.
2. Count the half squares: 2 half squares = 1 whole square.
3. Add to find the area: 2 + 1 = 3 square units.

Find the area of each figure.

1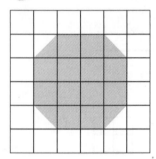

_____

3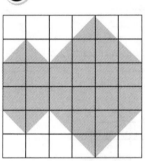

_____

5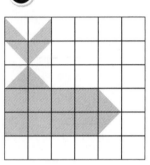

_____

2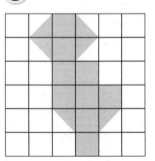

_____

4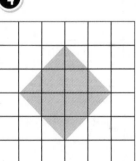

_____

6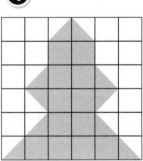

_____

Name _____ Date _____

Volume 1

Measurement

Find the volume by counting cubes.

 1

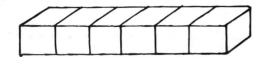

Volume = _____ cubic units

 2

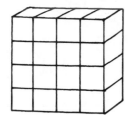

Volume = _____ cubic units

 3

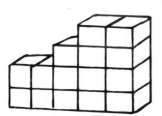

Volume = _____ cubic units

 4

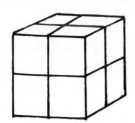

Volume = _____ cubic units

 5

Volume = _____ cubic units

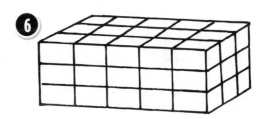 **6**

Volume = _____ cubic units

 7

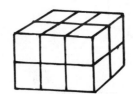

Volume = _____ cubic units

 8

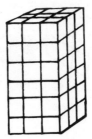

Volume = _____ cubic units

Volume 2

MEASUREMENT

You can find the volume by multiplying.
Volume = length x width x height

Multiply to find the volume of each figure.

 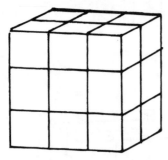

_____

 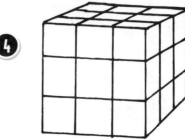

_____

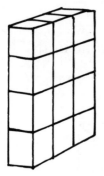 

_____

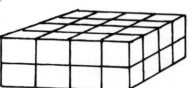

_____

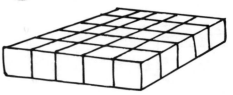

_____

Name _____   Date _____

Customary Units of Capacity

MEASUREMENT

The amount of liquid a container holds is called its **capacity**.

1 Tbsp	= 3 tsp	1 pt	= 2 c
1 oz	= 2 Tbsp	1 qt	= 2 pt
1 c	= 16 Tbsp	½ gal	= 2 qt
1 c	= 8 oz	1 gal	= 4 qt

Use the chart to find the equivalent measures.

1 5 Tbsp = _____ tsp

2 24 oz = _____ c

3 2 gal = _____ qt

4 8 c = _____ pt

5 8 oz = _____ Tbsp

6 24 pt = _____ qt

7 12 tsp = _____ Tbsp

8 12 oz = _____ Tbsp

9 24 oz = _____ c

10 6 Tbsp = _____ oz

11 12 pt = _____ qt

12 8 Tbsp = _____ c

13 32 oz = _____ c

14 7 qt = _____ gal

15 7 qt = _____ oz

16 6 pt = _____ qt

17 4 gal = _____ qt

18 6 c = _____ pt

19 14 Tbsp = _____ tsp

20 7 oz = _____ Tbsp

Standards-Based Math • 3–4 © 2004 Creative Teaching Press

Name _____ Date _____

Weight
MEASUREMENT

Weight measures how heavy an object is.

ounce (oz)
16 oz   = 1 pound (lb)
2000 lbs = 1 ton (T)

Use the chart to find the equivalent measures.

1 6000 lbs = _____ T

2 8 oz = _____ lb

3 1 T = _____ lbs

4 4 oz = _____ lb

5 32 oz = _____ lbs

6 6 T = _____ lbs

7 5 lbs = _____ oz

8 500 lbs = _____ T

9 2 lbs = _____ oz

10 4 T = _____ lbs

11 80 lbs = _____ oz

12 880 oz = _____ lbs

13 3 T = _____ oz

14 40 lbs = _____ oz

Which measure of weight would be most appropriate for each of the following items?

15 horse _____

16 slice of cheese _____

17 package of meat _____

18 a car _____

19 a person _____

20 can of soda _____

Standards-Based Math • 3–4 © 2004 Creative Teaching Press

Name _____ Date _____

Mass

MEASUREMENT

Mass measures the amount of matter in an object.

gram (g)
1000 grams   = 1 kilogram (kg)
1000 kilograms = 1 metric ton

Use the chart to find the equivalent measures.

1 5 kg = _____ g

2 500 g = _____ kg

3 6 metric tons = _____ g

4 1 metric ton = _____ g

5 17,000 g = _____ kg

6 1/2 metric ton = _____ kg

7 12 kg = _____ g

8 250 g = _____ kg

9 2500 kg = _____ metric tons

10 5500 g = _____ kg

Which measure of weight would be most appropriate for each of the following items?

11 a truck _____

12 a dollar bill _____

13 a baby _____

14 a piece of candy _____

15 a textbook _____

16 an elephant _____

Name _____ Date _____

Temperature–Fahrenheit

MEASUREMENT

In the customary system, temperature is measured in Fahrenheit.

95°F	70°F	32°F	10°F
It is hot outside.	It is warm outside.	Water freezes.	It is very cold outside.

Write the temperature shown on each thermometer.

1 Day 1   **2** Day 2   **3** Day 3   **4** Day 4

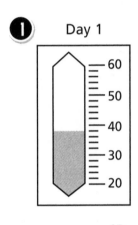   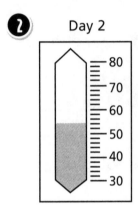   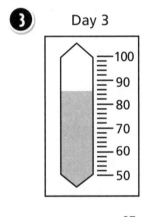   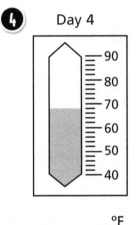

_____°F   _____°F   _____°F   _____°F

5 How much did the temperature change from Day 1 to Day 3? _____

Color each thermometer to show the temperature listed.

6   **7**   **8**   **9**

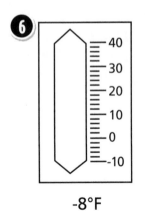   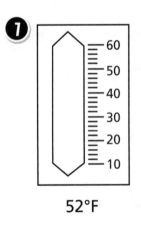   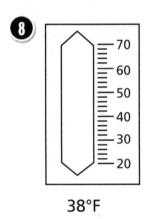   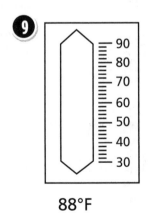

-8°F   52°F   38°F   88°F

Standards-Based Math • 3-4 © 2004 Creative Teaching Press

Name _____ Date _____

Temperature–Celsius

MEASUREMENT

In the metric system, the scale for measuring temperature is based on 100. It is called Celsius.

100°C	37°C	20°C	0°C
Water boils	Normal body temperature	Normal room temperature	Water freezes

Write the temperature shown on each thermometer.

1 Day 1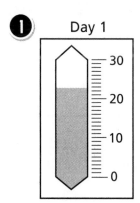

_____°C

2 Day 2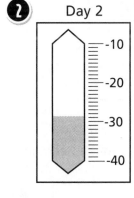

_____°C

3 Day 3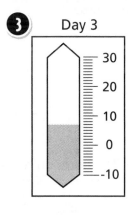

_____°C

4 Day 4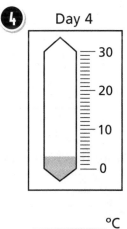

_____°C

Answer the following questions about Celsius with **yes** or **no.**

5 If the classroom were 23°C, would you need a coat? _____

6 You take your temperature. The thermometer reads 40°C.
Do you have a fever? _____

7 Would your ice melt in a room that was 0°C? _____

8 Could you boil noodles in a pot of water that was 102°C? _____

Name _____   Date _____

Money

MEASUREMENT

Write the amount of money shown.

①

②

③

_____   _____   _____

Draw the least number of bills and/or coins needed to make each amount shown.

④
$7.83

⑤
$2.52

⑥
$0.79

⑦
$2.13

⑧
$0.94

⑨
$5.47

Name _____ Date _____

Making Change

MEASUREMENT

Complete the table.

	Total Amount Owed	Amount Paid	Amount of Change Received
1	$25.46	$30.00	
2	$7.83	$10.00	
3	$11.46	$20.00	
4	$1.72	$5.00	
5		$15.00	$1.24
6	$0.47	$1.00	
7	$6.57		$3.43
8	$5.87	$6.00	
9	$29.72	$30.00	
10	$76.29	$100.00	

Standards-Based Math • 3-4 © 2004 Creative Teaching Press

Name _____ Date _____

Tally Chart

DATA ANALYSIS AND PROBABILITY

Use the tally chart to answer the questions.

Lunch Order	Tally
ham sandwich	IIII
grilled cheese	HHH III
salad	II
soup	HHH I

❶ How many people ordered soup for lunch? _____

❷ What was the most popular lunch item? _____

❸ How many people ordered ham sandwiches and salad? _____

❹ How many more people ordered grilled cheese than soup? _____

❺ Altogether, how many people ordered lunch? _____

Record the following information in the tally chart.

Student shirt colors: Green 7, Blue 8, Red 4, Pink 3, Orange 9

Student Shirt Colors	Tally

Standards-Based Math • 3–4 © 2004 Creative Teaching Press

Name _____ Date _____

Tables

Data Analysis and Probability

Use the table to answer the questions.

Grade 3 Pets	
Kind of Pet	Number of Students with Pet
Cat	12
Dog	9
Parakeet	6
Hamster	4
Fish	7

1 What is the most popular pet among third graders? _____

2 How many pets do the third graders have altogether? _____

3 What pet do the least number of third graders own? _____

4 How many parakeets and fish do the children own? _____

5 How many more cats do children own than fish? _____

6 How many four-legged animals do the third graders own? _____

7 How many more dogs are needed to equal the number of cats? _____

8 Order the third grader's pets from least to most owned.

_____

_____

_____

_____

Name _____ Date _____

Bar Graphs
DATA ANALYSIS AND PROBABILITY

Use the bar graph to answer the questions.

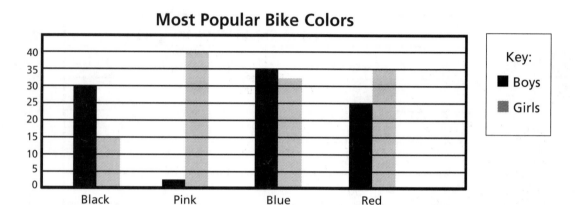

Most Popular Bike Colors

Key:
- ■ Boys
- ▨ Girls

1 What is the most popular bike color for boys? _____

2 What is the most popular bike color for girls? _____

3 Which bike color do boys and girls like almost the same? _____

4 How many girls like pink bikes? _____

5 How many boys and girls together like red bikes? _____

Create a bar graph using the information below.

Favorite vegetables of 4th graders: Carrots 20, Corn 40, Peas 10

Standards-Based Math • 3–4 © 2004 Creative Teaching Press

Name _____ Date _____

Pie Graphs

DATA ANALYSIS AND PROBABILITY

Use the pie graph to answer the questions.

Favorite School Subjects

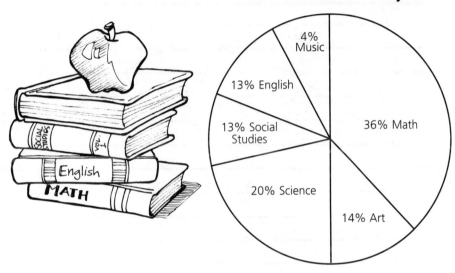

1 What subject received the most votes? _____

2 Which two subjects together received about half of the votes? _____

3 Which two subjects received the same percent of votes? _____

4 What percent of the vote did Art, Social Studies, and English receive? _____

5 Which subject received the smallest percent of the vote? _____

6 What percent more of the vote did Math receive than Music? _____

7 Did English and Social Studies together receive more, less, or the same percent of votes as Math? _____

8 Add all the percentages for each subject. What percent do you get? _____

9 What percent of students did not choose Math? _____

10 What percent of students did not choose Science or Social Studies? _____

Standards-Based Math • 3–4 © 2004 Creative Teaching Press

Name _____ Date _____

Line Graphs

DATA ANALYSIS AND PROBABILITY

Use the line graph to answer the questions.

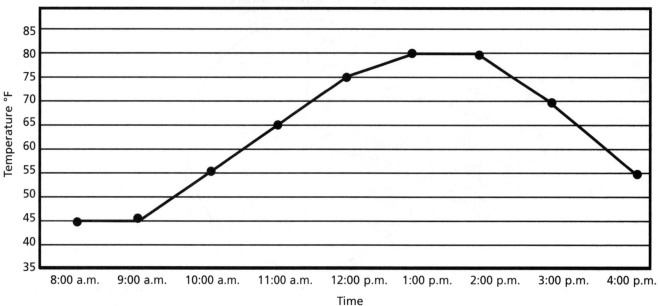

Temperature at School Playground

❶ What was the temperature at 8:00 a.m.? _____

❷ What was the highest temperature recorded on the playground? _____

❸ What is the title of this graph? _____

❹ At what time or times was the highest temperature recorded? _____

❺ At what time of day was the temperature 70° recorded? _____

❻ If students were outside at 9:00 a.m., what temperature was it? _____

❼ What was the temperature difference between 12:00 p.m.
   and 8:00 a.m.? _____

❽ At what time was the temperature higher, 3:00 p.m.
   or 12:00 p.m.? _____

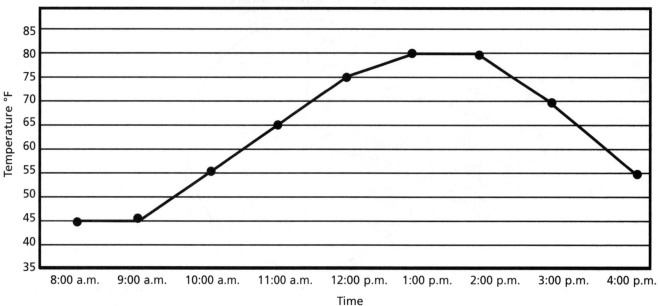

Standards-Based Math • 3–4 © 2004 Creative Teaching Press

Name _____ Date _____

Pictographs

DATA ANALYSIS AND PROBABILITY

Use the pictograph to answer the questions.

Number of Times Students Bounced a Basketball	
Kim	🏀 🏀 🏀
Mike	🏀 🏀 🏀 🏀
Alex	🏀 🏀
Jacob	🏀 🏀 🏀 🏀 🏀 🏀 🏀 🏀
Kevin	🏀 🏀 🏀 🏀
Denise	🏀 🏀 🏀 🏀 🏀 🏀 🏀
Key: 🏀 = 5 bounces	

1 What does this pictograph show? _____

2 What does the picture 🏀 represent? _____

3 How many times did Kevin bounce the ball? _____

4 Which person bounced the ball the most times? How many? _____

5 Which person bounced the ball the least times? How many? _____

6 Which two people bounced the ball the same number of times?
How many? _____

7 How many times did Jacob and Denise bounce the ball together? _____

8 How many more times did Denise bounce the ball than Alex? _____

Average

DATA ANALYSIS AND PROBABILITY

> **Average** is the sum total os a set of figures divided by the number of figures.
>
> String Lengths: 2 in., 4 in., 7 in., 3 in.
> 2 + 4 + 7 + 3 = 16
> 16 ÷ 4 = 4
>
> The average length of the strings is 4 in.

Find the average for each set of numbers.

1 3, 1, 4, 2, 5 _____

2 10, 20, 30 _____

3 12, 15, 18, 15 _____

4 17 cm, 36 cm, 46 cm _____

5 30, 17, 25 _____

6 1, 5, 10, 8 _____

7 8, 5, 4, 5, 3 _____

8 17, 8, 12, 5, 8, 4 _____

9 5, 7, 4, 4, 8, 2 _____

10 7, 8, 6, 3, 11, 9, 5 _____

Standards-Based Math • 3–4 © 2004 Creative Teaching Press

Mean, Median, Mode, Range

DATA ANALYSIS AND PROBABILITY

Mean: The mean is the average.

Median: The median is the middle number when data is ordered from least to greatest.

Mode: The mode is the value that occurs most often.

Range: The range is the difference between the greatest and the least numbers in a group.

Find the mean, median, mode, and range for each set of data.

Data	Mean	Median	Mode	Range
1 6, 5, 7, 6, 6				
2 10, 5, 3, 1, 1, 4, 11				
3 12, 13, 4, 1, 4, 2				
4 18, 11, 13, 5, 7, 5, 4				
5 20, 24, 3, 3, 5				
6 3, 1, 3, 2, 1, 3, 8				
7 13, 11, 13, 12, 9, 8, 4				
8 63, 64, 64, 65				

Likelihood of an Event

DATA ANALYSIS AND PROBABILITY

Write the letter of the bag that matches each probability statement.

A

B

C

D

E

_____ **1** It is unlikely that you will pick a dotted marble.

_____ **2** It is likely that you will pick a dotted marble.

_____ **3** It is certain that you will pick a dotted marble.

_____ **4** It is impossible for you to pick a dotted marble.

_____ **5** It is equally likely that you will pick a dotted marble.

Standards-Based Math • 3–4 © 2004 Creative Teaching Press

Determining Probability

DATA ANALYSIS AND PROBABILITY

$$\text{Probability} = \frac{\text{number of given events}}{\text{total possible events}}$$

The following types and numbers of cards are in a pile. Each time you choose one of the 20 cards, you return it to the pile so there are always 20 cards from which to choose.

  6   3   2   5   4

Determine the probability of each event.

1 What is the probability that you will pick a card with a circle on it? _____

2 What is the probability of picking a star card? _____

3 What is the probability of picking a square card? _____

4 What is the probability of picking a triangle card? _____

5 What is the probability of picking a diamond card? _____

6 What is the probability that you will pick a circle or a square card? _____

7 What is the probability that you will pick a diamond or a star card? _____

8 What is the probability that you will pick a circle, triangle, or star card? _____

Name _____   Date _____

Probability Picking

DATA ANALYSIS AND PROBABILITY

Solve the problems. Use the key to color the gumballs. Then answer the questions below.

25 = green   **60 = yellow**   **85 = blue**

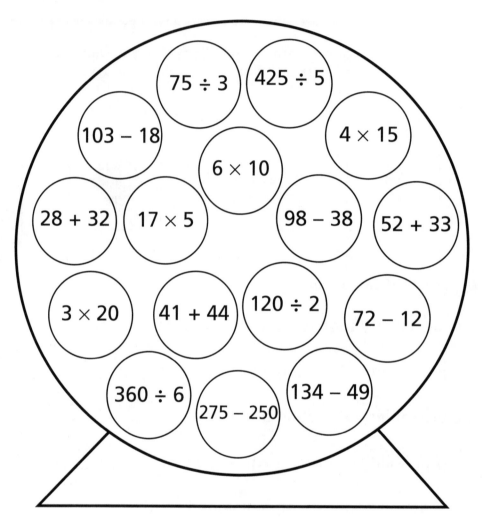

75 ÷ 3   425 ÷ 5

103 – 18   4 × 15

6 × 10

28 + 32   17 × 5   98 – 38   52 + 33

3 × 20   41 + 44   120 ÷ 2   72 – 12

360 ÷ 6   275 – 250   134 – 49

1 Is it likely or unlikely that you will get a yellow gumball? _____

2 Is it likely that you will get a red gumball? _____

3 Do you have a better chance of getting a green or a blue gumball?

_____

4 $\frac{1}{8}$ of the gumballs are what color? _____

Answer Key

Place Value (page 5)

1. ones
2. millions
3. hundred thousands
4. thousands
5. hundred thousands
6. ones
7. hundreds
8. ten thousands
9. tens
10. thousands
11. tens
12. ten thousands

Addition—No Regrouping (page 6)

1. 19	39	59	66	39
2. 89	68	58	79	54
3. 53	86	79	66	66
4. 67	77	86	67	49
5. 77	79	68	49	59

Score! (page 7)

98	149	150	98	99
97	97	96	75	74

Even—5
Odd—5

Addition—Regrouping (page 8)

1. 170	153	124	151	143
2. 121	122	167	113	161
3. 226	192	173	216	210
4. 247	186	212	230	301
5. 212	296	213	182	316

Three Addends (page 9)

1. 36	45	47	42	52
2. 52	72	87	71	101
3. 63	86	90	94	109
4. 57	70	77	80	105

Bowling for Numbers (page 10)

1. 45 + 25 + 30 + = 100
2. 75 + 15 + 10 + = 100
3. 35 + 15 + 50 + = 100
4. 45 + 45 + 10 + = 100
5. 50 + 40 + 10 + = 100
6. 55 + 25 + 20 + = 100

Word Problems 1 (page 11)

1. 82 points
2. 87 fourth-grade students
3. 5 more tickets
4. 78 pages
5. $18.00

Subtraction—No Regrouping (page 12)

1. 30	38	23	3	33
2. 20	32	4	52	102
3. 42	42	45	14	105
4. 223	424	512	622	232
5. 742	524	816	542	910

Take a Moonwalk (page 13)

Subtraction—Regrouping (page 14)

1. 19	45	47	8	23
2. 89	89	49	28	59
3. 178	369	86	47	29
4. 113	58	156	529	16
5. 177	26	43	245	448

Number, Please (page 15)

328	467	237	328
Zanville	Carlstown	Linden	Zanville
178	178	237	467
Jackson	Jackson	Linden	Carlstown

Multiplication—No Regrouping (page 20)

1. 48	22	39	63	82
2. 44	66	70	66	68
3. 88	88	20	26	86
4. 90	84	64	48	44
5. 36	28	93	99	96

Subtraction Soup (page 16)

1. 48	32	27	132
2. 71	18	47	89
3. 64	18	77	162

Multiplication—Regrouping (page 21)

1. 52	34	65	72	50
2. 72	75	154	108	70
3. 220	128	87	104	90
4. 135	112	192	378	110
5. 184	238	376	486	385

Word Problems 2 (page 17)

1. 58 people
2. 37 coins
3. 320 photos
4. 60 magazine subscriptions
5. 52 pieces of paper

Two-Digit Multiplication—No Regrouping (page 22)

1. 132	143	169	273	504
2. 264	506	370	651	748
3. 516	308	840	403	946
4. 605	200	1,024	525	924

Multiplication 0–5 (page 18)

1. 0	6	24	40
2. 3	9	9	14
3. 16	21	15	8
4. 12	15	44	45
5. 8	27	28	20
6. 16	24	40	55
7. 18	0	16	30
8. 28	24	25	21

Dragon Digits (page 23)

1. 968; 276; 176
2. 140; 1,056; 533
3. 736; 165; 451
4. 168; 299; 903

The fourth dragon should be colored.

Two-Digit Multiplication—Regrouping (page 24)

1. 182	192	195	273	540
2. 432	644	592	1,702	814
3. 688	812	1,470	455	1,932
4. 2,145	1,196	1,216	700	2,016
5. 792	1,752	1,512	2,072	1,408

Multiplication 6–12 (page 19)

1. 24	14	24	45
2. 18	54	28	32
3. 63	56	30	42
4. 16	36	48	36
5. 56	80	81	99
6. 60	63	96	90
7. 84	64	49	108
8. 66	27	121	144

Multiplying for Medals (page 25)

1. 92; gold	92; gold	72; silver
2. 70; silver	57; silver	
3. 48; bronze	33; bronze	65; silver
4. 96; gold	100; gold	

Three-Digit Multiplication—Regrouping (page 26)				
1. 860	428	1,565	738	2,905
2. 1,554	1,488	965	1,444	1,638
3. 2,538	398	1,248	1,239	1,944
4. 1,575	1,416	1,056	900	2,140

Four-Digit Multiplication—Regrouping (page 27)				
1. 7,245	3,814	12,852	13,230	14,445
2. 11,276	24,992	23,433	17,444	11,068
3. 12,711	15,672	36,348	9,166	9,944
4. 15,375	21,708	11,916	13,180	12,120

Word Problems 3 (page 28)

1. 64 balloons
2. $1.75
3. 10 cousins
4. 80 daffodil bulbs
5. 3 rolls of string

Division 0–5 (page 29)				
1. 7	4	2	6	3
2. 5	1	0	4	0
3. 3	6	3	2	5
4. 6	1	7	4	2
5. 3	9	8	3	7
6. 2	4	5	4	3

Skyscraper Stories (page 30)

```
            2       2
        4       6       9
        8       3       8
    8           8       2
    9           3       8
    5           2       3
    9           4       9
```

The Sears Tower is 110 stories tall.

Two-Digit Division—No Remainders (page 31)			
1. 23	13	11	12
2. 14	21	11	23
3. 16	13	19	13
4. 18	12	17	39
5. 24	29	15	26
6. 14	14	11	17

Three-Digit Division—No Remainders (page 32)			
1. 213	213	122	242
2. 173	132	26	22
3. 28	36	18	21
4. 58	61	42	64
5. 72	96	84	97

One-Digit Division—Remainders (page 33)			
1. 2R1	1R2	1R3	2R1
2. 4R1	2R2	1R1	1R2
3. 1R3	3R1	1R1	1R1
4. 1R2	1R1	1R2	1R1
5. 1R3	1R4	1R2	2R1

Rise and Shine (page 34)			
1. 6R1	5R3	7R1	9R5
2. 3R3	9R3	7R5	6R5

Three-Digit Division—Remainders (page 35)			
1. 115R1	126R2	63R1	208R2
2. 97R3	40R4	281R2	67R1
3. 242R1	152R1	47R8	180R2
4. 122R7	76R2	73R4	167R4
5. 257R2	56R3	84R3	135R5

Apple Seed Solutions (page 36)

John Chapman

```
10   15   8    14
3    8    1    16   13   1    14
```

Word Problems 4 (page 37)

1. 28 people
2. 26 gallons
3. 12 books
4. $24.00
5. 78 shells

Adding Fractions with Like Denominators (page 38)

1. $3/4$ $2/3$ $3/5$
2. $4/6$ $5/6$ $2/4$
3. $5/9$ $6/7$ $4/5$
4. $7/10$ $5/8$ $4/6$
5. $6/6$ $2/4$ $6/7$

Subtracting Fractions with Like Denominators (page 39)

1. $2/4$ $2/5$ $3/6$
2. $1/5$ $2/7$ $2/6$
3. $4/9$ $2/8$ $3/7$
4. $5/6$ $1/9$ $2/8$
5. $3/5$ $3/10$ $1/7$

Adding Fractions with Unlike Denominators (page 40)

1. $3/6$ $4/6$ $3/4$
2. $5/12$ $5/8$ $8/9$
3. $5/6$ $11/12$ $7/10$
4. $4/4$ $11/15$ $11/14$

Subtracting Fractions with Unlike Denominators (page 41)

1. $2/8$ $5/10$ $3/6$
2. $5/8$ $1/6$ $5/12$
3. $3/10$ $6/10$ $1/6$
4. $5/12$ $1/10$ $7/15$

Reducing Fractions (page 42)

1. $1/2$ $1/2$ $1/3$ $1/3$
2. $1/3$ $1/3$ $2/3$ $1/3$
3. $1/4$ $3/4$ $2/3$ $3/8$
4. $2/7$ $1/3$ $4/9$ $2/5$
5. $1/5$ $1/2$ $1/3$ $1/2$
6. $1/7$ $3/4$ $3/16$ $2/9$

Word Problems 5 (page 43)

1. 14 people
2. 2 sisters
3. 3 blocks
4. 1 1/6 cups
5. 3/8 of her water bottle

Repeated Patterns (page 44)

1. □ ▭ □

2. B D C

3. ○ ☆ □

4. snap clap tap

5. × = +

6. ⛵ 🚗 🚗

Number Patterns (page 45)

1. 12, 14, 16, 20
2. 75, 70, 65, 60
3. 15, 18, 21, 24
4. 27, 32, 37, 42
5. 32, 64, 128, 256
6. 42, 40, 38, 36
7. 50, 60, 70, 80
8. 100, 75, 50, 25
9. 20, 21, 25, 26
10. 46, 47, 94, 95

Number Chart Patterns (page 46)

These numbers should be **green**:
5, 10, 15, 20, 25, 30, 35, 40, 45, 50,
55, 60, 70, 75, 80, 85, 90, 95, 100

These numbers should be **blue**:
2, 4, 6, 8, 10, 12, 14, 16, 18, 20, 22, 24, 26, 28, 30, 32, 34, 36,
38, 40, 42, 44, 46, 48, 50, 52, 54, 56, 58, 60, 62, 64, 66, 68,
70, 72, 74, 76, 78, 80, 82, 84, 86, 88, 90, 92, 94, 96, 98, 100

These numbers should be **red**:
2, 3, 5, 7, 11, 13, 17, 19, 23, 29, 31, 37, 41, 43, 47, 53, 59, 61,
67, 71, 73, 79, 83, 89, 97

These numbers should have a **purple X** over them:
4, 8, 12, 16, 20, 24, 28, 32, 36, 40, 44, 48, 52, 56, 60, 64, 68, 72, 76, 80, 84, 88, 92, 96, 100

Positive and Negative Integers (page 47)

1. <	6. <	11. >
2. <	7. <	12. >
3. >	8. <	13. =
4. >	9. <	14. <
5. <	10. <	15. <

Order Properties (page 48)

1. T	6. T
2. F	7. F
3. T	8. T
4. T	9. F
5. F	10. F

Equality Properties (page 49)

1. 5	6. 1
2. 3	7. 8
3. 10	8. 7
4. 7	9. 56
5. 4	10. 100

Missing Addends and Subtrahends (page 50)

1. 6	7. 14
2. 15	8. 12
3. 8	9. 9
4. 24	10. 23
5. 31	11. 7
6. 10	12. 41

A deck of cards

Missing Factors and Divisors (page 51)

1. 4	7. 12
2. 3	8. 7
3. 11	9. 8
4. 5	10. 6
5. 20	11. 13
6. 9	12. 18

a garbage truck

Function Tables (page 52)

1.	6	4.	7
	9		14
	10		5
	13		8
	16		4
	12		3

2.	14	5.	12
	28		18
	7		32
	0		23
	77		20
	42		15

3.	6	6.	16
	3		64
	9		24
	17		40
	13		56
	5		72

What's the Rule? (page 53)

1. + 6
2. x 5
3. ÷ 4
4. x 7
5. - 3
6. ÷ 2

Find the Rule (page 54)

1.	x 9	4.	– 2
	27		13
	6		7
	1		18

2.	+ 9	5.	÷ 6
	1		10
	14		36
	9		1

3.	+ 4	6.	– 9
	5		0
	20		20
	15		12

Writing Expressions (page 55)

1. $3 + r$
2. $10 - n$
3. $4 \times c$
4. $p \div 3$

Order of Operations (page 56)

1. 22
2. 17
3. 2
4. 15
5. 5
6. 18
7. 18
8. 12
9. 2
10. 20
11. 12
12. 54
13. 25
14. 24

Writing Equations with Variables (page 57)

1. $3 + n = 7$
2. $39 - c = 28$
3. $4 \times b = 28$
4. $m \div 8 = 8$

Writing and Solving Equations (page 58)

1. $125 + c = 300$
$c = 175$ cans
2. $3 \times m = 36$
$m = 12$ inches
3. $7 \times r = 49$
$r = 7$ rows
4. $251 - b = 114$
$b = 137$ books
5. $27 - s = 12$
$s = 15$ students

Solving Equations 1 (page 59)

1. 5
2. 19
3. 3
4. 6
5. 2
6. 8
7. 21
8. 11
9. 44
10. 3
11. 23
12. 36
13. 7
14. 4
15. 9
16. 57
17. 5
18. 13
19. 45
20. 21

Solving Equations 2 (page 60)

1. 7
2. 4
3. 15
4. 6
5. 216
6. 1
7. 4
8. 7
9. 56
10. 6
11. 7
12. 7
13. 7
14. 2
15. 7
16. 21
17. 11
18. 4
19. 5
20. 3

Formulas (page 61)

1. $2(55) + 2(23) = 156$ yards
2. $22 \times 15 = 330$ square feet
3. $2(6) + 2(4) = 20$ feet
4. $16 \times 15 = 240$ square feet

Find That Item (page 62)

1. (0, 2)
2. (2, 1)
3. (5 ,5)
4. (1, 6)
5. (1, 3)
6. (4, 1)
7. (7, 4)
8. (3, 5)
9. (3, 3)
10. (6, 2)

Mystery Picture (page 63)

a house

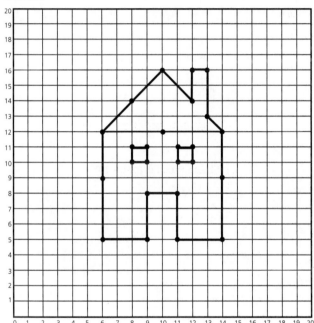

Language of Lines (page 64)

1. f
2. a
3. e
4. b
5. g
6. c
7. d

Connect the Line Segments (page 65)

a star

Identify the Angles (page 66)

1. acute
2. acute
3. obtuse
4. obtuse
5. straight
6. right
7. right
8. straight

Identify the Shapes (page 67)

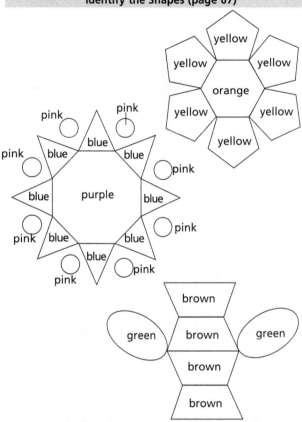

Polygon or Not? (page 68)

Polygons
B
D
F
J
L

Not Polygons
A
C
E
G
H
I
K

Draw That Shape (page 69)

Trapezoid

Square

Hexagon

Parallelogram

Rectangle

Circle

Rhombus

Quadrilateral

Oval

Triangle

Pentagon

Polygon

quadrilateral
parallelogram
rhombus
square
polygon

Naming The Triangles (page 70)

1. equilateral
2. scalene
3. equilateral
4. isosceles
5. obtuse
6. acute
7. right
8. obtuse

Label the Circle (page 71)

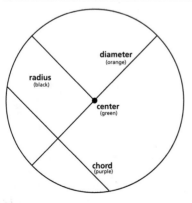

Congruent of Not? (page 72)

1. yes
2. no
3. no
4. yes
5. yes
6. yes
7. no
8. no
9. yes
10. no

Slides, Flips, and Turns (page 73)

1. turn
2. slide
3. turn
4. flip
5. flip
6. flip

Shape Puzzler (page 74)

1. Q
2. A
3. H
4. M
5. L
6. I

Make Similar Figures (page 75)

Answers may vary, but drawings should reflect a figure of the same shape but not necessarily the same size.

Lines of Symmetry (page 76)

1. no
2. yes
3. no

4.

5.

6.

7.

8.

9.

10.

11.

12.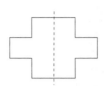

How Many Lines? (page 77)

1. 0
2. 4
3. 4
4. more than 4
5. 3
6. 1
7. 0
8. more than 4
9. 0
10. 1

Parts of Solid Figures (page 78)

Square   5, 5, 8,
Triangle 5, 6, 9
Rectangle 6, 8, 12
Hexagon 8, 12, 18
Triangle 4, 4, 6

What Am I? (page 79)

1. cylinder
2. sphere
3. cone
4. cone
5. cylinder
6. cube
7. cube
8. sphere
9. pyramid
10. cube

Identifying Bases (page 80)

1. circle
2. square
3. circle
4. triangle
5. square
6. circle

Draw That Figure (page 81)

Prism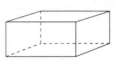

Sphere

Cylinder

Cone

Cube

Pyramid

cube, prism

Solid Figure Riddle (page 82)

Footsteps

Similar or Congruent? (page 83)

1. similar
2. congruent
3. congruent
4. similar
5. congruent
6. similar
7. congruent
8. similar
9. congruent
10. congruent

What Time Is It? (page 84)

Minutes will vary slightly.

1. 12:25
2. 2:36
3. 11:08
4. 9:54
5. 7:13
6. 5:06
7. 6:30
8. 8:44
9. 3:00

Figure the Time (page 85)

1. 8:15 a.m.
2. 3:00 p.m.
3. 12:30 p.m.
4. 1 hour and 50 minutes
5. 3:39 p.m.
6. 15 minutes

Calendar Days (page 86)

1. November 15
2. November 5
3. November 25
4. 15 days
5. November 24
6. November 21
7. November 17

Nonstandard Units of Measurement (page 87)

Answers will vary.

Customary Measurement (page 88)

1. 4 3/4 inches
2. 1/2 inch
3. 3 1/4 inches
4. 2 inches
5. 1 1/2 inches
6. 4 inches

Metric Measurement (page 89)

1. 4 1/2 cm
2. 9 cm
3. 6 cm
4. 10 1/2 cm
5. 3 cm
6. 8 cm

Computing Customary Units (page 90)

1. 2
2. 5,280
3. 84
4. 48
5. 9
6. 3,520
7. 3
8. 5
9. 27
10. 8
11. inches
12. feet
13. yards
14. miles
15. inches or feet
16. inches
17. miles
18. yards

Computing Metric Units (page 91)

1. 100
2. 4
3. 3000
4. 7000
5. 2000
6. 2
7. 4000
8. 30
9. 10
10. 20

millimeter
centimeter
decimeter
meter
kilometer

Perimeter (page 92)

1. 52
2. 40
3. 42
4. 27.6
5. 55
6. 18
7. 36
8. 53

Area 1 (page 93)

1. 9 square units
2. 12 square units
3. 8 square units
4. 5 square units
5. 10 square units
6. 12 square units
7. 12 square units
8. 6 square units
9. 8 square units
10. 10 square units
11. 4 square units
12. 24 square units

Area 2 (page 94)

1. 14 square units
2. 12 square units
3. 22 square units
4. 8 square units
5. 12 square units
6. 17 square units

Volume 1 (page 95)

1. 6
2. 16
3. 15
4. 8
5. 28
6. 45
7. 12
8. 54

Volume 2 (page 96)

1. $2 \times 3 \times 3 = 18$ cubic units
2. $3 \times 4 \times 1 = 12$ cubic units
3. $4 \times 4 \times 2 = 32$ cubic units
4. $3 \times 3 \times 3 = 27$ cubic units
5. $2 \times 3 \times 6 = 36$ cubic units
6. $5 \times 5 \times 1 = 25$ cubic units

Customary Units of Capacity (page 97)

1. 15
2. 3
3. 8
4. 4
5. 16
6. 12
7. 4
8. 24
9. 3
10. 3
11. 6
12. $^1/_2$
13. 4
14. 1.75 or $1^3/_4$
15. 224
16. 3
17. 16
18. 3
19. 42
20. 14

Weight (page 98)

1. 3
2. 1/2
3. 2000
4. 1/4
5. 2
6. 12,000
7. 80
8. 1/4
9. 32
10. 8000
11. 1,280
12. 55
13. 96,000
14. 640
15. ton
16. ounce
17. pound
18. ton
19. pound
20. ounce

Mass (page 99)

1. 5000
2. 1/2
3. 6,000,000
4. 1,000,000
5. 17
6. 500
7. 12,000
8. 1/4
9. 2 1/2
10. 5 1/2
11. metric ton
12. gram
13. kilogram
14. gram
15. kilogram
16. metric ton

Temperature—Fahrenheit (page 100)

1. 38
2. 54
3. 86
4. 68
5. 48

6.
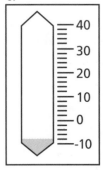

7.
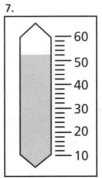

8.
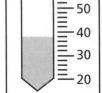

9.
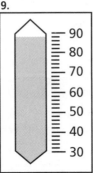

Temperature—Celsius (page 101)

1. 23
2. -29
3. 7
4. 3
5. no
6. yes
7. no
8. yes

Money (page 102)

1. $2.73
2. $5.57
3. $0.98 or 98¢

4.

5.

6.

7.

8.

9.

Student Shirt Colors	Tally
Green	ⵑⵑⵑ II
Blue	ⵑⵑⵑ III
Red	IIII
Pink	III
Orange	ⵑⵑⵑ IIII

Tables (page 105)

1. cats
2. 38
3. hamsters
4. 13
5. 5
6. 25
7. 3
8. hamsters, parakeets, fish, dogs, cats

Bar Graphs (page 106)

1. blue
2. pink
3. blue
4. 40
5. 60

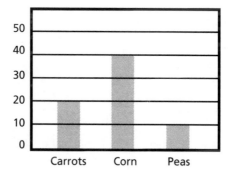

Making Change (page 103)

1. $4.54
2. $2.17
3. $8.54
4. $3.28
5. $13.76
6. $0.53
7. $10.00
8. $0.13
9. $0.28
10. $23.71

Tally Chart (page 104)

1. 6
2. grilled cheese
3. 6
4. 2
5. 20

Pie Graphs (page 107)

1. Math
2. Math, Art
3. English, Social Studies
4. 40%
5. Music
6. 32%
7. less
8. 100%
9. 64%
10. 67%

Line Graphs (page 108)

1. 45°
2. 80°
3. Temperature at School Playground
4. 1:00 p.m. and 2:00 p.m.
5. 3:00 p.m.
6. 45°
7. 30°
8. 12:00 p.m.

Pictographs (page 109)

1. Number of times students bounced a basketball
2. 5 bounces
3. 20 times
4. Jacob; 40 times
5. Alex; 10 times
6. Mike and Kevin; 20 times
7. 75 times
8. 25 times

Average (page 110)

1. 3
2. 20
3. 15
4. 33 cm
5. 24
6. 6
7. 5
8. 9
9. 5
10. 7

Mean, Median, Mode, Range (page 111)

1. 6; 6; 6; 2
2. 5; 4; 1; 10
3. 6; 4; 4; 12
4. 9; 7; 5; 14
5. 11; 5; 3; 21
6. 3; 3; 3; 7
7. 10; 11; 13; 9
8. 64; 64; 64; 2

Likelihood of an Event (page 112)

1. D
2. E
3. A
4. B
5. C

Determining Probability (page 113)

1. 6/20
2. 4/20
3. 2/20
4. 3/20
5. 5/20
6. 8/20
7. 9/20
8. 13/20

Probability Picking (page 114)

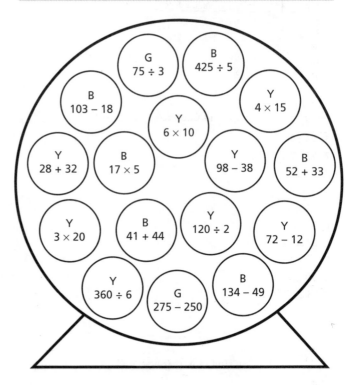

1. likely
2. no
3. blue
4. green